13 Reasons To Doubt:

Essays from the writers of Skeptic Ink

Edited by
Edward K Clint,
Jonathan MS Pearce,
and
Beth Ann Erickson

Acknowledgements

We would like to thank Jonathan MS Pearce and Beth Ann Erickson for their diligent work editing this volume and the Skeptic Ink'ers themselves for their contributions, particularly Russell Blackford, Maria Maltseva and Rebecca Bradley for their tireless copyediting. We also want to express our gratitude for James Randi, Ray Hyman, Michael Shermer, Joe Nickell, and D.J. Grothe for their inspiring work in skepticism, along with Hypatia of Alexandria (our patron secular icon), Carl Sagan and Paul Kurtz.

If you like what you read in this book then visit us at: http://www.skepticink.com

Contents

SKEPTIC INK NETWORK

Introduction

This book is an introduction to the voices of Skeptic Ink, a web salon promoting science, reason, and a reality-based worldview. We created Skeptic Ink in the summer of 2012 in response to a clear need for improved support for open, critical discourse within and without the secular and skeptical movements. Writers at Skeptic Ink speak to a broad range of topics: religion, politics, science, paranormal claims, feminism, philosophy, secularism, and much more. This is consonant with our conviction that skepticism should be a part of daily life, and not merely a periodic activity. It is our bread and butter, not our rainy day at the bowling alley (as much fun as that can be).

The two dozen writers are themselves quite diverse, hailing from Australia, the United States, Canada, Britain, Columbia, Ireland, South Africa, Russia, and Iran. We have tried and succeeded to some degree to develop a secular world perspective. They are professionals from such domains as writing and publishing, public advocacy, anthropology,

1

archaeology, law, and philosophy. Included are an ex-Muslim, ex-Jew, a former preacher, an ex-fundamentalist Christian, and several former believers in the paranormal. Making the transition from true believer to doubter is often personally difficult, but it leaves one with a respect for the positive transformative power of skepticism, as well as the ability to empathize with people who still suffer in thrall. Both are key to our mission of promoting skepticism as an ideal which can embiggen us all. Does the internet need one more skeptic borough? In a word, yes.

The Easy and Hard Problems of Skepticism

The aims and purposes of organized skepticism may be separated into two related but distinct groups of problems. The first we will call the "easy" problem, even though it is quite difficult in execution: exposing frauds such as psychics, consumer protection against snake oil such as homeopathic products, and advocating for good science. These efforts meet with limited success, as fraudulent psychics and vendors nonetheless proliferate. We refer to them as "easy" because demonstrating the point is straightforward: psychics always fail testing, as does homeopathic "medicine." Moreover, it is easy by comparison. The hard problem of skepticism is helping people to internalize and apply the powerful tools of critical thinking and skepticism in their daily lives, to their own thoughts and opinions. A brief glance at history shows how difficult this is and how rarely it happens: Ayn Rand championed objectivity and founded a philosophy based on "pure reason," but nonetheless became what Michael Shermer has termed the leader of a personality cult. Aristotle discovered and described what we now call logic, but he also held false ideas that logic and empirical evidence did not support, such as that women have more teeth than men. Albert Einstein was a consummate educated scientist who showed real independent thinking. In spite of that, he invented the cosmological constant only so that the universe would conform to his personal expectations of universal rectitude (to his credit, he later called this the biggest mistake of his career).

Skepticism, in this latter sense, is hard. It is not reliably remedied by education, intelligence, or familiarity with skepticism itself. Some of the reasons it is so difficult are discussed in Part 1. This is disturbing in a world in which humans have ever-expanding technological power to diminish or promote human flourishing and to cause or prevent global ecological devastation. It is equally disturbing at the smallest scales, when children die because their parents believe they can treat an easily curable illness with prayer or New Age snake oil. While some of the psychological causes of the hard problem of skepticism are understood, solutions are not obvious. The answer to the "why" of Skeptic Ink is that solving both groups of problems is incredibly important, and organized skepticism needs all the help it can get. Now on to the "how." We look to answers from the one human endeavor which has, so far, coped with the problem of human frailty the best, science.

Unblinding Them with Science

The scientific enterprise has out-competed every way-of-knowing humanity has ever come up with. Arguably, this is because it employs an antagonistic open discourse and peer-review system which leverages the power and biases of other minds to conduct error-checking. That is to say, it can take a weakness (personal bias) and turn it into a positive good (peer review, research finding replication). The public nature of this error-checking also provides a powerful incentive for people to self-check in order to head off potential embarrassment. The system is far from perfect, but when it comes to increasing human understanding of the world, science dramatically outperforms any other system ever devised. We do not suggest that skepticism can or should function exactly like science, but that the critical error-checking elements can be reformulated for general public consumption.

There are three aims of Skeptic Ink which can help address the hard problem of skepticism. The first is promotion of the virtue of critical thinking. The virtue of critical thinking specifies the structure of useful discourse. It privileges objectivity and dulls the sting of motivated dissent by making it mundane. The second is the promotion of open and public discourse about important issues. As in science, a marketplace of ideas where anyone can participate is likely to fare better in the long run. These first two depend upon the healthy functioning of the last: antagonism. Granted, we mean civil and productive antagonism. Consider the Olympics, the rise of democracy over tyranny, and the success of competitive economic markets over non-competitive markets (as in capitalistic monopolies or state-controlled monopolies). Or just think of science. In each, we can see how the spirit of competition can bring out the best in humanity. Ideas should compete with each other in a fair arena such that the best of them rise to the top.

In a nutshell, Skeptic Ink exists to help promote skepticism in general: across all topics and in day to day life. We aim to make critical discourse more accessible, more mundane, and more basic to society. That said, we tend to comment on a narrow range of topics that happen to reflect our particular bases of expertise and experience where much good can be done. This book is thusly divided into three parts. Part One explores the intersection of skepticism, belief, and science and some of the reasons why the hard problem of skepticism is so hard. Part Two includes essays on religion, which remains one of the most powerful sociopolitical factors of global events. This part includes discussion of some of the latest intellectual skirmishes between religionists and atheists. The final section is about skepticism and atheism today, including modern perspectives on the atheism movement and the path to skeptical atheism for some Skeptic Ink writers, no doubt mirroring the experiences of millions of others.

Edward K Clint
Los Angeles, June 2014

SKEPTIC INK NETWORK

6

A Brief History of Doubt: Great Skeptics from Antiquity to the Renaissance
Peter Ferguson

Humanity's thirst for knowledge, combined with our need to be able to explain our surroundings, constitutes one of our greatest characteristics—but also one of our greatest downfalls. Historically, whenever we were unable to explain a phenomenon, we did not have the capacity to leave it unanswered. It was in our nature to supply an explanation, to provide order to the perceived chaos. But as we lacked the necessary knowledge to acquire an accurate conclusion, mankind resorted to inventions. Fairy-tales and myths were told to explain our surroundings, natural resources, weather, and our origin; in fact, almost everything was given a mythological back-story. Hence gods and religion were invented. However, for as long as religion has existed, so too have people who were willing to expose its falsehoods and seek the truth behind the universe and ourselves. This chapter will chronicle such people from antiquity to the Renaissance. The focus will be on Europe as it is the European philosophers and skeptics who have influenced the Western world and the three Abrahamic religions the most. Beginning in Greece, we will visit the early philosophers before departing for the Roman world where we shall witness the emergence of skepticism and science before finishing in the Middle Ages with Islamic intellectualism.

The Golden Age of Greece, 500-300 BC, produced some of the greatest thinkers of humanity. It is during this period that we find some of the earliest doubters and skeptics of religion and its claims. The citizens of Greece were religiously conservative and the charge of atheism was very serious. However, being accused of being an atheist did not mean the person did not believe in a god or gods, it simply meant that the person refused to adhere to the gods recognized by the state. For instance, Socrates was accused of being an atheist despite him claiming he was divinely inspired, yet he was still sentenced to death; others were forced into exile. So to challenge the gods was to risk one's own life. Yet Ancient Greece was littered with such men who were willing to do just that.

Protagoras[1] (490-420 BC) was a pre-Socratic philosopher whose texts we have lost bar a few fragments and a number of citations in other authors' works. In one such lost work, *On The Gods,* Protagoras proffers an agnostic philosophy: "Concerning the gods, I have no means of knowing whether they exist or not or of what sort they may be, because of the obscurity of the subject, and the brevity of human life." Such a statement was bold for its era and according to ancient authors it caused some turmoil. Cicero and Diogenes Laertius both report that Protagoras was exiled from Athens and his books burnt. However, their accounts are unreliable as they were written centuries after the event and several modern scholars have questioned the veracity of these claims. Although it is not possible to prove whether Protagoras was indeed exiled and his books burned, the simple fact remains that ancient authors considered such actions plausible due to the nature of the sentiment.

Yet this is not Protagoras' most famous or most controversial statement; he also posited a form of relativism which was radical for the age: "man is the measure of all things: of things which are, that they are, and of things which are not, that they are not." The original context of this expression has been lost and the phrase comes to us via later authors who quote Protagoras. Due to the lack of

---

[1] Not to be confused with Pythagoras of Pythagorean Theorem fame.

context, the meaning is not entirely explicit and is somewhat open to interpretation. However, the fact that this quote was deemed worthy of quotation by numerous ancient authors indicates that it was important and highly regarded. It was one of the earliest suggestions that knowledge was subjective, based on the observations of humanity and there was no such thing as objective truth. Cultural values and morals were not gifted from the gods to humans but were judged through human experience, and each person was the purveyor of their own moral authority. Protagoras was one of the earliest intellectuals to question the existence of the gods and the idea of an absolute truth and morality.

Democritus (460-370 BC), a contemporary of Protagoras, also divorced the gods from human morality. He based his morals on nature and believed there were no supernatural or divine influences. Democritus suggested that goodness derived from practice and self-discipline rather than some innate characteristic of humanity. Wicked people should be avoided as in their presence one was likely to be wicked too. Freedom from fear was imperative to achieving happiness, a condition which can be arrived at through balance and moderation as opposed to hedonism. Despite his radical notion of morality it is not what Democritus is best known for. Alongside his mentor, Leucippus, he developed one of the earliest atomic theories. They hypothesized that everything consisted of minute indivisible particles called atoms (*atoma*). These atoms were indestructible, infinite in number, in constant motion, and differed in shape and size. This theory is remarkably close to the modern understanding of atomic theory; in fact, it was not until the 18th century that the theory was advanced in any significant manner.

A student of the Democritean school, Epicurus (341-270 BC), did develop upon the theory slightly. He suggested that everything is the result of the motion and collisions of atoms: these atoms travelled through empty space with no predetermined plan colliding and ensnaring each other, thus creating everything. His theory excluded and negated the need for a supernatural engineer. Although Epicurus never denied the existence of any gods, he did question their presence in the human domain. He believed that the gods were simply not interested in the affairs of humans and that they did not interfere with the mortal realm in any manner. The most famous quote traditionally attributed to him is on the problem of evil: how can an omnipotent god and evil exist simultaneously?

> Is God willing to prevent evil, but not able?
> Then he is not omnipotent.
> Is he able, but not willing?
> Then he is malevolent.
> Is he both able and willing?
> Then whence cometh evil?
> Is he neither able nor willing?
> Then why call him God?

Epicurus also did not believe in the afterlife, as a materialist; he did not think the soul could exist outside of the human body: "For when we are, death is not: and when death is, we are not." It was due to this philosophy that he emphasized happiness and friendship throughout his life. He had a large group of friends including slaves and women whom he treated as equals. He went as far as allowing them into his school which would have been quite shocking for its era.

When we cross over to the Roman world, we find it is very similar to that of the Greek. The gods are almost identical but have Romanized names. Any impiety was met with the charge of atheism, whether or not the person actually was an atheist. Christians were often accused of being atheists despite their theism simply because they failed to recognize the official Roman gods. However, unlike in Greece, religion in the Roman world was less conservative and impiety was rarely ever punished. Christians were, of course, persecuted but that was because they professed a different god to the Romans. However, impiety such as deism, agnosticism and atheism went relatively unpunished. We find that many of the Greek doubters and skeptics had huge influence on their Roman counterparts.

Lucretius (99-55 BC) was a follower of Epicurus and Democritus. He wrote an epic poem *De Rerum Natura* (*On the Nature of Things*), divided into six books, which espoused the Epicurean philosophy. Lucretius wrote the poem in an effort to explain the Epicurean principle of atomism, physics, the development of the world, and the nature of celestial and terrestrial phenomena to the Roman people. Lucretius details how superstition is a curse and leads to immoral acts. He argues that mankind should rid itself of the fear of the unknown:

> ... the mind, divorced from anxiety and fear, may enjoy a feeling of contentment. And so the nature of the body evidently is such that it needs few things, namely those which banish pain and, in so doing, succeed in bestowing pleasures in plenty [...] this terrifying darkness that enshrouds the mind must be dispelled not by the sun's rays and the dazzling darts of day, but by the study of the superficial aspect and underlying principle of nature.

Lucretius rejects the idea of a supernatural designer: "the world was by no means created for us by divine agency." He describes the movements of heavenly bodies and rejects their divine status. An account of the origin of the world is given and the development of humanity and society is detailed. In the final book, Lucretius explains that phenomena such as earthquakes, thunder, lightning, and volcanoes are natural and should not be attributed to gods: "terrestrial and celestial phenomena which, when observed by mortals, make them perplexed and panic-stricken, and abuse their minds with dread of the gods." Lucretius sustained Epicureanism and developed upon it. He spread its tenets to the Roman people and influenced many of its greatest thinkers including Cicero and Vergil.

Pliny the Elder (23-79 AD) also studied and tried to explain natural phenomena. He wrote a fantastic treatise *Naturalis Historiae* (*The Natural History*) which was a vast encyclopedia that attempted to detail all the ancient knowledge of the time. It was extremely successful and became the base model for later encyclopedias. In his *The Natural History*, Pliny has a section on "The Search for God." In this section, Pliny never outright rejects the idea of a god but he does proffer an agnostic philosophy and rejects many of the claims made in the name of religion. He argues that if everybody's version of god existed then there would be more gods than men, and this simply cannot be the case, especially since many nations worship animals and other creatures he deems repulsive. He also finds it ridiculous to believe that, if there is a god or gods, they pay any heed to human affairs; this places Pliny firmly in the realm of deism if not agnosticism:

> But it is ridiculous to suppose, that the great head of all things, whatever it be, pays any regard to human affairs. Can we believe, or rather can there be any doubt, that it is not polluted by such a disagreeable and complicated office?

Pliny also rejects any idea of an afterlife; he believes it is just the wishful thinking of mortal men who fail to come to terms with their own demise. Pliny was a naturalist who compiled all the information which was available to him into a single book which became the greatest compendium of knowledge to have ever existed at that time.

Despite the fact that the people I have discussed thus far have all been innovative thinkers who rejected the prevailing theories about religion, cosmology, philosophy etc., they still had critics of their own. Sextus Empiricus (160-210 AD) was a physician and a philosopher who could be considered a skeptic in the modern sense. Sextus was a Pyrrhonian skeptic: he belonged to a school of thought that was named after Pyrrho, a Greek philosopher (360-270 BC). Pyrrhonian skeptics would refrain from making any assertions until enough information was accrued. They would argue both sides of an issue and attempt to make both propositions as strong as possible but if one argument failed to be more convincing that its counterpart then judgment was suspended until one side presented itself with a stronger case. It is for this reason that many Pyrrhonians considered many other schools of thought to be dogmatic, even forward thinking philosophies such as Epicureanism, because they made positive claims which Pyrrhonians considered non-evident, and such assertion without evidence was dogmatism. A number of Sextus' works survive: his *Pyrrhōneioi hypotypōseis* (*Outlines of Pyrrhonism*) and several treatises leveled against mathematicians, arithmeticians, musicians, astrologers, etc. Sextus' work influenced many great philosophers in later centuries, such as Friedrich Hegel, David Hume, and Renee Descartes.

Due to the patriarchal nature of ancient society, each of my examples of a skeptic has been male. There was, however, a female skeptic whose fame has survived to this day.

It was extremely rare for a woman to receive any form of education or training and their societal roles were strictly defined. This meant that the chances of a female thinker of the caliber listed in this chapter developing in such a stifling environment were almost nonexistent. Yet, we do find one such person in Egypt: Hypatia. Her birth date is uncertain but it was circa 370 AD and she died 414 AD. Educated in mathematics, philosophy, and astronomy by her father, she headed a Platonist school in Alexandria which was unheard of for a female. She became a symbol of science and learning and many flocked to her to study under her tutelage. After years of teaching in Alexandria she became embroiled in a political dispute between the city's Bishop, St Cyril, and its prefect (civil leader) Orestes. It was an ancient church-state separation clash as Orestes believed that St Cyril was encroaching upon his civic duties.

As the dispute escalated it got violent as supporters of both clashed in the streets of Alexandria. Hypatia became a target as she was a friend of Orestes and her teachings, such as her philosophy and mathematics, were attributed to paganism. Her influence, scholarship and scientific knowledge were a threat to the Christian supporters of St Cyril and she was abducted and brought to a church where she was killed.[2] Some scholars argue that her death marked the downfall of intellectualism in Alexandria. Many scholars abandoned Alexandria as they feared persecution because their teachings were being targeted for accusations of heresy and paganism.

From the 5th century onwards, religious skeptics become scarce. Catholicism had become the official religion of the Roman Empire and the persecution of pagans was well underway. Criticism of Catholicism and its claims was dealt with far more fiercely than it was under paganism which meant that religious skeptics and critics were unlikely to speak out. There was also an intellectual vacuum during this

---

[2] I thoroughly recommend the excellent film *Agora* which does a great job of dramatizing her life.

period: philosophy, literature and scientific endeavors were replaced by biblical study. The abandonment of such important and thought-provoking subjects coupled with the persecution of heretics meant there was little chance a religious skeptic would surface during this period. However, there was a reemergence of religious skepticism during the Islamic Golden Age which lasted from the 8th until the mid-13th century. During this period there was a focus on intellectualism and the Arab world became a center for science, literature, philosophy, mathematics, and medicine.

Muhammad ibn Zakariyā Rāzī (865-95 AD) was a physician, alchemist, chemist, mathematician, and scholar. He made numerous advances in each of these fields and authored over 200 works, two of which were criticisms of religion: *Fī al-Nubuwwāt* (*On Prophecies*) and *Fī Ḥiyal al-Mutanabbīn* (*On the Tricks of False Prophets*). Rāzī was particularly critical of religions which claimed to have been revealed through prophecy:

> ... [God] should not set some individuals over others, and there should be between them neither rivalry nor disagreement which would bring them to perdition.

> On what ground do you deem it necessary that God should single out certain individuals [by giving them prophecy], that he should set them up above other people, that he should appoint them to be the people's guides, and make people dependent upon them?

Rāzī also witnessed the violence which emerged from religious belief and the unwillingness of religious adherents to question their beliefs:

... there would be a universal disaster and they would perish in the mutual hostilities and fighting. Indeed, many people have perished in this way, as we can see.

If the people of this religion are asked about the proof for the soundness of their religion, they flare up, get angry and spill the blood of whoever confronts them with this question. They forbid rational speculation, and strive to kill their adversaries. This is why truth became thoroughly silenced and concealed.

He also recognized the power of indoctrination. He believed the people were originally deceived by religious leaders and the same leaders continue the deception:

... as a result of [religious people] being long accustomed to their religious denomination, as days passed and it became a habit. Because they were deluded by the beards of the goats, who sit in ranks in their councils, straining their throats in recounting lies, senseless myths.

Although Rāzī was viewed as a heretic by his contemporaries he was able to live his life unmolested under the relatively liberal Islamic rule. This is partly due to the fact that Islamic society was more tolerant than the Christian society at this time and Rāzī's work was highly valued. He was one of the greatest physicians of his age and he made numerous discoveries. He was the first to differentiate between measles and smallpox, and he discovered chemicals such as kerosene.

Abul ʿAla Al-Maʿarri (973-1058 AD) was a blind poet, philosopher, and religious critic, and most notably a rationalist who valued reason over dogma and superstition. He recognized that religion was completely fabricated and rejected any concept of divinity. He also believed that religion benefited nobody but those in charge and their priests. His poems are small and easily digestible, and accurately and concisely convey his views on religion.

Creation Reveals A Lack of Sense
You said, "A wise one created us ";
That may be true, we would agree.
"Outside of time and space," you postulated.
Then why not say at once that you
Propound a mystery immense
Which tells us of our lack of sense?

The Two Universal Sects
They all err—Moslems, Jews,
Christians, and Zoroastrians:
Humanity follows two world-wide sects:
One, man intelligent without religion,
The second, religious without intellect.

Death's Debt is Paid in Full
Death's debt is then and there
Paid down by dying men;
But it is a promise bare
That they shall rise again.

What is Religion?
What is religion? A maid kept close that no eye may view her;
The price of her wedding gifts and dowry baffles the wooer.
Of all the goodly doctrine that I from the pulpit heard
My heart has never accepted so much as a single word.

Fools Awake!
O fools, awake! The rites ye sacred hold
Are but a cheat contrived by men of old
Who lusted after wealth and gained their lust
And died in baseness—and their law is dust.

In his poetry Al-Ma'arri tackles many facets of religious belief: he dismisses an intelligent creator, disregards any notion of an afterlife, highlights the falsehood of rituals, ridicules the intelligence of all the religious regardless of specific faith, and argues that the priests profit from their deceit. Al-Ma'arri is also the only person discussed in this chapter who verges on atheism. Due to the lack of scientific knowledge it was impossible for people to conclude there was no god. In fact, we find rather few atheists in the modern sense until the 20th century, when there was finally enough data to discount the possibility of a god. However, by dismissing the concept of a creator in one of his poems, Al-Ma'arri has come pretty close to the premise of atheism.

The people chronicled in this chapter are some of the most famous and revered of their time. They authored some of the greatest works and developed revolutionary philosophies, even formulating some early scientific theories. The respective periods in which each individual lived were all heavily religious, when many were persecuted for questioning the prevailing religion. There was little to no knowledge of the origin of humanity nor was there any understanding of natural phenomena such as volcanoes and earthquakes; everything was attributed to invented gods. Such an environment stifled independent thinking and suppressed freedom of inquiry. Despite this, these individuals advanced complex ideas and concepts. The method which these people utilized to construct such ideas was simple: asking questions. They refused to tacitly accept the predominant philosophical and religious models. By doing so, agnosticism was proffered, a position many hold today quite easily; but for its time it was quite a novel and contentious position. The origin of morality was discussed, initially thought to stem from the gods; alternative ideas were put forth including moral relativism by Protagoras and a form of deterministic morality by Democritus. Such ideas divorced morality from the gods, and people began to see morality as an innate characteristic of humans rather than some divine gift. More

18

impressively, Democritus developed the atomic theory, simply by asking the question "What are we are made of?" This theory was very advanced for its time and unfortunately wasn't significantly built upon for over 2000 years. By rejecting religious beliefs, many began to see the negative effects religion had on society. Lucretius recognized how the inability to understand natural phenomena terrified people and that superstition led people to commit immoral acts. Muhammad ibn Zakariyā Rāzī highlighted the power of indoctrination and violent reaction people had when their beliefs are being questioned.

Much of what was discussed may seem eerily familiar and that is because we are still having these same debates today. The nature of morality is still being discussed with many people still attributing our ethical nature to god; some go as far as stating that non-religious people lack any form of morality. The origin of the universe and humanity is still debated despite our ever-expanding compendium of knowledge regarding the composition of the universe. There is still the violent streak among some religionists who refuse to question their religion; some even outlaw it under the threat of severe penalties. Superstition still encourages people to commit immoral acts; we still witness witches being burnt and albinos being hunted to be cannibalized. All this is because the idea of questioning everything has not permeated the entirety of society. Good ideas are capable of withstanding the rigors of scrutiny while bad ideas are tossed aside. Without doubt, bad and harmful ideas persist. It is only by continuing to question everything, including, if not especially, ourselves, in the same vein as those detailed in this chapter can we hope to rid ourselves of persistent irrationality and indoctrination.

Bibliography

Bailey, A. (2002). *Sextus Empiricus and Pyrrhonean Scepticism*. Oxford: Oxford University Press.

Butterworth, C. E. ed. (1990). *The Political Aspects of Islamic Philosophy: Essays in Honor of Muhsin S.Mahdi.* Princeton: Princeton University Press.

Cartledge, P. (1999). *Democritus.* New York: Routledge.

Esolen, A. M. (1995) *Lucretius On the Nature of Things.* Baltimore: John Hopkins University Press.

Guthrie, W. K. C. (1977). *The Sophists.* New York: Cambridge University Press.

Hecht, J. M. (2004). *Doubt: A History: The Great Doubters and Their Legacy of Innovation from Socrates and Jesus to Thomas Jefferson and Emily Dickinson.* New York: Harper Collins.

Jones, H. (1989) *The Epicurean Tradition.* London: Routledge.

Kerferd, G.B. (1981). *The Sophistic Movement.* Cambridge: Cambridge University Press

Rist, J.M. (1972). *Epicurus: An Introduction.* London: Cambridge University Press.

Skepticism in an Age of Ideology
Russell Blackford

The Passing of an Age of Faith

I'd like you to think back about five hundred years. More specifically, consider European civilization around the year 1500, a time at the end of the medieval era or, if you prefer, the beginning of Western modernity. Much that was plausible *then* seems extraordinary to us *now*; much that was then unthinkable is now very thinkable indeed—even intellectually attractive. Much has happened to alter humanity's understanding of the world.

The celebrated Canadian philosopher Charles Taylor has carried out a similar exercise to what I'm asking of you, though his relates specifically to the thinkability of atheism. I'll want us to range a bit more broadly, but let's begin with Taylor. His 2007 book, *A Secular Age*, discusses how much changed during the past few centuries to enable a transition from a society where belief in God was essentially unchallenged to the current situation in Western societies where it is, as Taylor puts it, "understood to be one option among others, and frequently not the easiest to embrace" (2007, 3). (Note that Taylor is himself a Christian believer and his book is far from being a defense of atheism or an attack on religion.)

Taylor emphasizes that there was no fully-developed non-religious alternative available in Europe around 1500. That is, no comprehensive (or even somewhat less than comprehensive) secular worldview was on the table as an option. Moreover, he suggests, there were three features of European society that made atheism unthinkable. First, the natural world was seen as testifying to divine purpose and

21

action, both in its appearance of order and in occasional "acts of God" such as plagues, disasters, and years of exceptional fertility. Second, social life, with all its levels, practices, and constituent components and associations, was seen as underpinned by God. Everyday life was pervaded by prayer, ritual, and worship. Third, there was a strong sense of living in an enchanted cosmos, full of miraculous agents, objects, and powers (Taylor 2007, 25–28; compare Blackford and Schüklenk 2013, 192–193).

This seems like a reasonable summary of why the sixteenth century was still an age of religious faith. And despite the intellectual and artistic stirrings of the Renaissance, it was an age of *unchallenged* faith. In 1500, some of the pressures that would lead a decade or two later to the Protestant Reformation were in evidence, but they involved matters of church organization, anxieties about worldliness and corruption, and specific doctrinal details (as they now seem to us, though of course many of these "details" were of great importance to the political and theological antagonists of the time). There was little or no serious wish on anyone's part to overthrow Christianity itself or to cast doubt on God's existence. Well-developed forms of philosophical atheism still lay well in the future, and great reformers such as Martin Luther and John Calvin never imagined that they were involved in a process that would ultimately make atheistic philosophies more socially acceptable.

If we follow an account such as Taylor's, the rise of atheism required a desanctification of political and social life, new understandings of the natural and cultural worlds, and a progressive disenchantment of the cosmos. Many factors probably contributed to these, perhaps including increased urbanization and even tendencies within medieval theology itself, but certainly including the Reformation with its fatal challenge to the authority of the Catholic Church. The

Reformation also made possible the bloodthirsty wars of religion of the sixteenth and seventeenth centuries, particularly the catastrophic Thirty Years' War of 1618-1648, which involved most of Europe in one way or another and left millions dead. Such horrors produced an understandable backlash, as seventeenth-century thinkers began to devise entirely secular theories of state power.

The rise of science during this same period provided alternative explanations of many events previously assigned to supernatural causes (compare Haack 2007, 268-272). Gradually, the successes of science undermined the pre-existing explanatory systems. The problem for these systems, which were inextricably entangled with religious traditions, hierarchies, and doctrines, was not merely that science disagreed with them on specific points—important as this was. Perhaps more importantly, the old explanatory systems were revealed as premature and—by the newly emerging standards—poorly evidenced. In turn, this cast doubt on the intellectual standing of religious organizations and traditions, ancient books, folklore, and other supposed sources of knowledge.

I began by stating that I wanted us to think more broadly than about belief in God. It was not only God-belief that came rather easily to sixteenth-century Europeans. All sorts of ideas that seem bizarre to scientifically educated people today were commonplace. Much efficacy was attributed to spirits, miracles, and magic. By contrast, we now explain the workings of the natural world in scientific terms. Even where we still lack full scientific explanations, as with the emergence of life from non-life (a quite different issue from the diversity and intricate functioning of life forms), many of us, quite justifiably, do not take non-scientific explanations seriously.

Some Cheers for Science and Skepticism

Times have changed, and ours is no longer an age of faith. The sciences do not, admittedly, provide us with a final and comprehensive picture of the world, but they have discovered much that is likely to stand the test of time. Many scientific findings are now so well evidenced that we can think of them as robust—it is most unlikely that they will ever be overturned, however much might eventually be added to our understanding. For example, there is no prospect that scientists will change their minds and conclude that the sun revolves around the earth, that illnesses are caused by miasmas or evil spirits, or that the myriad forms of life we see all around us sprang into existence simultaneously some 6000 years ago.

Nonetheless, many people seem to resist the scientific way of looking at the world, as if they were still living at the dawn of Western modernity. Many extraordinary claims continue to be popular; new ones appear each year, originating from many sources. I take contemporary scientific skepticism to be, in response, a certain (perhaps exasperated) suspicion about extraordinary claims.

Consider one of my favorite examples, one with an ancient pedigree in both Western and Eastern thought: reincarnation. Scientific skeptics will take a suspicious attitude to the claim that reincarnation is a real phenomenon. If they treat the claim at all seriously, they will subject it to searching and harsh scrutiny. The justification for this is simply that reincarnation does not fit readily into the picture of the world built up by science over the past four or five centuries. If the evidence for it became so compelling that we concluded that it must be a real phenomenon, we would need to revise much of our scientific understanding. Putting this yet another way, the whole record of science in

24

establishing its (admittedly provisional and incomplete) world picture stands in the balance against any evidence for reincarnation.

That is not a reason to close our minds completely against reincarnation claims. But we can make a reasonable judgment that they are most unlikely to turn out to be true. We can legitimately treat them as extraordinary and ask for exceptionally powerful evidence before we believe them.

Not all claims that we should now regard as extraordinary are anomalous in this strong sense—i.e., they are not so astounding that our whole scientific picture, as developed so far, would need to be revised if they were true. Indeed, some very broad and non-specific religious claims, including bare claims about the existence of a divine creator, need not outright contradict any particular scientific findings. The problem for religious institutions is that they have lost much of their authority and prestige in the last few centuries, even though many churches have revised their doctrines to avoid the most direct kinds of inconsistency with science.

Science continues to challenge many strongly held and emotionally salient beliefs about the world, including beliefs about human beings and human societies. In some cases, as I've suggested, it tends to reveal entire explanatory systems as premature and poorly evidenced. In others, specific claims about the world simply do not stand up to testing: see, for example, the often heartfelt, but multiply debunked, allegations made by anti-vaccination campaigners. Still others merit suspicion because they contradict well-established findings from empirical disciplines, such as history, that lie outside of what is usually regarded in English-speaking countries as "science," or "the sciences." For example, we should now regard the claims of Holocaust denialists as extraordinary, even though a focus on them may not be central to the mission of scientific skepticism.

We might once have had reason to question claims about the Holocaust, given the sheer scale and horror of the crimes involved, as well as much historical experience with atrocity propaganda. Compare the extravagant lies that were spread about the German military during World War I. For example, British propagandists alleged that the Germans had a factory to boil down the corpses of their soldiers to extract human fat for use in household and military products (Marlin 2002, 71-76). Since that time, there have been innumerable examples of atrocity allegations that turned out to be dubious or just plain wrong.

There is no doubt, however, that the Nazi Holocaust, including the murders of almost six million Jews, actually took place. Given the case that was built by investigators immediately after World War II, and by many historians since, we actually have the remarkable evidence needed to believe in the occurrence of something as monstrous as the Holocaust. Someone who now denies those events does not deserve the honorable title of "skeptic." Such a person is a denialist (and it should come as no surprise that Holocaust denialists are often genuinely anti-Semitic, blinded by hate and ideology).

As a generalization, however, it remains true that atrocity narratives are often exaggerated, false, or even fraudulent. At the same time, it is difficult to refute or confirm the scary stories so often employed by our political leaders as justifications to embark on a war (or continue fighting one). Given the human tendency to believe the worst of supposed or potential enemies, the ease with which atrocity narratives can be fabricated and spread, and the motivation for very powerful players, such as national governments, to do exactly that, we should approach these stories with suspicion.

Propaganda, Ideology, and Belief

Though ours is not an age of faith, it is often identified as an age of propaganda (e.g. Pratkanis and Aronson 1992, xiii; Marlin 2002, 13). Indeed, the mainstream and social media constantly flood us with one-sided, emotionally manipulative, and often dishonest efforts to influence our opinions. "Everyday," Pratkanis and Aronson wrote in the early 1990s, "we are bombarded with one persuasive communication after another. These appeals persuade not through the give-and-take of argument and debate but through manipulation of symbols and of our most basic human emotions" (1992, 5-6). Sometimes, as with much histrionic journalism, the aim may be pure sensationalism. No specific effect on our beliefs or behaviors may be intended, as long as all the drama prompts us to buy newspapers, watch current affairs programs, or click on populist blogs. Yet the consequences may go much further in distorting public perceptions, provoking overreactions, and threatening our liberties (see Radford 2003, 147-177, 221-260).

Our age is also an age of ideology. While the authority and prestige accorded to religion may have been reduced by hundreds of years of progress in secular thought and rational inquiry, there remains a tendency for human beings to generate, seek out, and promote comprehensive worldviews—and then cling to them desperately. And so, the past two centuries have been dominated by a succession of ideologies and isms that teach us in detail how to live our lives and exercise whatever power we might have over others. These worldviews invariably go far beyond what is justified by good evidence.

At this point, we might take note that many strong claims made by social scientists or humanities scholars, even perfectly reputable ones, have nothing like the evidential

27

backing possessed by say, heliocentric theory in astronomy, evolutionary theory in biology, or the knowledge of the Holocaust possessed by historians. The social sciences and humanities are enormously important, but they often overreach. They are rife with contested findings and often sharply divided along political lines; many individuals are motivated by contestable moral or ideological commitments. At best these fields of learning are in a pre-scientific state. At their worst, they can display intellectual dishonesty, motivated reasoning, deliberate obscurity, and an unbecoming tribalism.

That is not a warrant for hyperskepticism about the humanities and social sciences. It is merely a warrant for *reasonable* levels of skepticism. Clearly much honest, brilliant, and rigorous work goes on in all these fields. I am not writing a diatribe against honest scholars, and I must acknowledge that I am, myself, attached to my local university's School of Humanities and Social Science. But we do have reason for caution. That is especially so when we're confronted with scholarly opinions that run up against common experience, call for drastic action, make a fetish of verbal obscurity, or rely on evidence that is open to multiple interpretations.

There is another aspect of all this. What confronts us as we go about our lives, trying to make the best sense of the world that we can, is often not some claim that contradicts our scientific picture of the universe as a whole, or even some specific (and robust) empirical finding. Rather, the evidence for and against many claims is often simply too murky for an intellectually honest and responsible person to reach any confident conclusion. In those circumstances, it may often be wiser to suspend judgment than rely on hunches, emotional reactions, and personal or political loyalties.

When It's Smart (and Honest) to Suspend Judgment

Getting to the truth of some issue that we find important can be maddeningly frustrating. Often, we must first settle the truth of numerous *other* claims before we can even know what the evidence really amounts to. Given our limited resources, as individuals and collectively, it may, in a particular instance, be almost impossible to establish any truth about which we should honestly be confident. In some circumstances, critical evidence is lost or unobtainable or just plain ambiguous. If so, there might be no practical investigations that could tell us decisively (or even with a high degree of confidence) what the truth actually is.

I am not arguing here for any kind of radical, comprehensive skepticism. I am most definitely not denying that there are objective truths. Nor am I suggesting, only a bit less radically, that we can never be justifiably confident about anything at all. Rather, my claim is one that I think we *can* be confident about: in a very wide range of situations, we are not in a position to draw conclusions with a confidence that we could justify to, let's say, a rational and open-minded stranger.

Sometimes we have no satisfactory option but to bet on our largely-intuitive assessments of the probabilities. In many other cases, though, there may be no pressing reason to do anything but suspend judgment until more compelling evidence becomes available—if that ever happens. But in any event, why do we so often feel the pressure to take a stance, perhaps even one that is backed by strong emotions, on issues where we can't be justifiably confident one way or the other? Why, for example, should I form a view of my own about a high-profile criminal trial when I have nothing like the advantages of the jury in seeing and hearing

29

the full evidence admitted at trial? Why be emotionally invested in a particular outcome, when the jury is almost certainly better placed than I am to make a decision? (Even if some evidence is kept from the jury, though available to me through the mass media, I should remember that evidence is often inadmissible at trial for good reason. That is, it may be highly prejudicial but of little probative value.)

We should say more often, if only to ourselves: "This issue is conceptually difficult and the evidence seems murky; I'm not sure what the truth is here." That might not feel satisfying or win you friends, but it will show a certain kind of honesty.

The Burdens of Intellectual Honesty

There is, of course, a danger that even true and beneficial messages will encounter public cynicism in this age of propaganda and ideology. It is widely known that many of the messages we receive are misleading and manipulative, and that many people take positions based on their ideological commitments. Once we learn this lesson, we may feel more impelled to think for ourselves. Fine, but we might then find ourselves taking agnostic positions even on questions where the facts are fairly much established and the answer ought to be clear to well-informed people. Unfortunately, it can be difficult to know what to think or to assess who does or does not deserve our trust. The mainstream media can make this worse by giving time or column inches to cranks and crackpots, treating them with the same respect as genuine experts. Don't even start me on the way social media have often been used to undermine sound and genuine science when it comes to such button-pressing issues as biological evolution and climate change.

Again, I reject any sort of radical skepticism. Some individuals do, indeed, merit our trust, at least on matters where they have expertise and are not likely to be blinded by ideological commitments. But *which* individuals? How do we answer the bald question, "Who is an expert?" (Pigliucci 2010, 280), and the more general question of who, expert or not, can be trusted to show good sense on a particular issue? Difficult though this is, it's not an altogether hopeless task. Let's be honest with ourselves. Most of us can already identify people whom we believe when it comes to subjects beyond our everyday knowledge because they seem (and probably are) genuinely expert. But if we do some self-searching, most of us can identify others whom we're inclined to believe because doing so is emotionally comfortable or confirms our biases, or because their words support some cause to which we've become attached.

Intellectual honesty about that distinction would take us a long way, and we can go somewhat further. Following a study by Alvin Goldman, Massimo Pigliucci describes how we can assess the credibility of alleged experts, examining 1. the formal cogency of their arguments, 2. whether these "experts" are supported by other experts in the field, 3. whether there is independent evidence of their expertise, 4. what biases they appear to have, and 5. what track record they have established (Pigliucci 2010, 291-301). Particularly when we are dealing with a mature scientific field with a substantial body of robust findings, it may be quite apparent who is an expert and who is not.

All the same, it will frequently make sense to tread warily. In an age of propaganda and ideology, we continue to hear messages that we ought to doubt and to be confronted by issues where the most honest approach is to suspend judgment. Even where it is appropriate to reach a conclusion one way or another, often we should hold to it with a degree of tentativeness, remaining open to revision of our beliefs should the balance of evidence shift.

We can also take steps to avoid rewarding propagandists and ideologues, and to resist perpetuating the cycle (as must happen if we all become propagandists and ideologues ourselves). That does not mean having no commitments, but our commitments should include a range of pro-truth values. Here, I am asking a lot—and some of it may be unpalatable. I'm asking for a commitment to represent opposing positions fairly; a commitment not to engage in weakly justified character attacks, or in attempts to browbeat, bully, gang up upon, or personally vilify opponents; and a commitment to look for the strengths as well as the weaknesses in positions that you dislike. Will you make a commitment like that?

Intellectual progress requires a habit of mind where we stop and consider what might be the *strongest* arguments against our own positions or in favor of positions that we oppose. It requires a willingness to change our minds, an appreciation of reasonable, good-faith opponents, and an openness to facts that might be surprising and inconvenient.

The Specter of Quietism

One concern about all this might be a fear that it will lead to quietism. If we are unwilling to pursue urgent causes until we have certainty that we are correct, might we fail to take part in important social debates and political movements? Consider such dramatic social evils as chattel slavery and racial segregation. Surely these are cases where there was an urgent need for political mobilization that could not wait for philosophical or scientific certainty or even for the best arguments to be devised. If we always have to wait for the perfect argument, perhaps we'll hold back at times when passionate, unalloyed commitment is needed.

Surely there is *something* to this concern. Where there is much at stake in the sense of enormous human suffering, we may need to act forcefully and immediately, rather than worry about which are the strongest arguments against our positions or whether our rationales for action are exactly the correct ones. But once again, I have not argued that we can never be confident about anything. Nor do I suggest that reasonable confidence has anything to do with the concept of certainty. In the circumstances of, say, South Africa's apartheid policy there was plenty of reason for confidence that the policy was an evil one, meriting vehement opposition.

Unlike, say, the ancient Pyrrhonian skeptics, I am not advocating quietism. But nor will I excuse destructive behavior in pursuit of dubious causes. Generally speaking, intellectual honesty, well-grounded skepticism, and a policy of opposing propaganda and ideology will not prevent us from opposing well-evidenced evils, but sometimes they actually will justify caution or restraint. They may quite rightly incline us to suspend judgment about the latest heinous allegations (perhaps against a political opponent), or to keep cool heads in the midst of the latest moral panic. With many issues, an honest assessment would show that the situation is far from clear, that even our allies are motivated more by ideology than evidence, and that some of our opponents may be decent people with legitimate points that are worth consideration.

Once again, you may be able to identify causes where the facts are all on your side and some drastic action definitely needs to be taken to prevent large-scale suffering or oppression. Certainly, there are causes like that, and you needn't respond to them with quietism. By all means, speak up and play your part. But I'm betting that most of us can also identify pet causes where, come to think of it, the facts are less clear and we are inclined to take our stances more because of tribal identification, confirmation of our biases, emotional associations, and so on.

Conclusion

Many claims—whether about ghosts, magic, miracles, reincarnation, or other extraordinary phenomena—should be viewed with suspicion because they sit badly with our picture of the world from modern science. Scientific findings are provisional, and much remains to be discovered. Nonetheless, in practice many of the main findings of the sciences are now so well-evidenced that they are unlikely ever to be overturned. We know much that can be regarded, for practical purposes, as established and solid.

If we are intellectually honest, we will find that many of the old systems of thought and explanation have been revealed as premature and lacking in any convincing evidence. Though they may be elegant, comforting, and impressively encyclopedic, these systems are now long past their use-by dates. We can admire the art, architecture, philosophy, and other achievements that they inspired, but there is no good reason to take them seriously as guides for living our lives or understanding the world around us. That applies, I submit, to all the world's religions.

I am not arguing for any kind of radical or comprehensive skepticism, and there is much that we should be confident about. Nonetheless, we live in an age of propaganda and ideology, and many people feel the attractions of comprehensive belief systems, even though these remain poorly evidenced and premature. The decline of religion in the twentieth century was accompanied by a rise of secular ideologies that were equally dubious and often highly destructive.

Intellectual honesty and fairness to others can make difficult demands on us, such as that we suspend judgment on many emotionally compelling issues. It may not come easily to us, but I suggest we refuse to follow the dubious wisdom of our particular tribes, or to adopt beliefs that please others whose approval we want. Instead, we should look for relevant evidence. In the large class of cases where we simply don't know the answer to some question, why not be open about it? In very many situations, the simple words *I don't know* are not a sign of weakness or culpable ignorance, but of honesty and frankness. We should hear—and speak—those words more often.

We should be more honest with others, and above all with ourselves, about how little we actually do know and how incomplete any defensible understanding of the world will inevitably be. Conversely, we should recognize how premature and overreaching any comprehensive worldview will be. Here and in other publications, I've offered advice on countering propaganda and ideology. There are many things we can do, such as listening more closely to reasonable, good-faith critics of our positions. Most importantly, though, we need a mindset that is more accepting of uncertainty, ambiguity, and an element of incompleteness in our understanding of the world. Are we up to that challenge?

References

Blackford, Russell and Udo Schüklenk. (2013). *50 Great Myths About Atheism*. Malden, MA: Wiley-Blackwell.

Haack, Susan. 2007. *Defending Science—within Reason: Between Scientism and Cynicism.* Amherst, NY: Prometheus.

Marlin, Randal. (2002). *Propaganda and the Ethics of Persuasion.* Peterborough, Ontario, and Orchard Park, NY: Broadview Press.

Pigliucci, Massimo. (2010). *Nonsense on Stilts: How to Tell Science from Bunk.* Chicago: University of Chicago Press.

Pratkanis, Anthony R. and Elliot Aronson. (1992). *Age of Propaganda: The Everyday Use and Abuse of Persuasion.* New York: WH. Freeman.

Radford, Benjamin. (2003). *Media Mythmakers: How Journalists, Activists, and Advertisers Mislead Us.* Amherst, NY: Prometheus.

Taylor, Charles. (2007). *A Secular Age.* Cambridge, MA: Harvard University Press.

Are You A Skeptic?

Maria Maltseva

Before I get into the substance of this article, I have to admit, I hate it when people apply self-congratulatory labels to themselves, and "skeptic" often seems to be such a label. It becomes even more ridiculous when the person doing so isn't the least bit skeptical and completely unaware of the various biases he or she possesses. Given all this, I have to say that I am definitely not a skeptic in any formal sense, but I do think critically about many issues, I question authority, I evaluate the credibility of the experts I rely on for information, and I try to avoid groupthink. As to whether I'm successful in all these endeavors; well, I'm skeptical.

In truth, I stumbled on skepticism by accident. I was looking for something akin to a luxury booze cruise, and I found one sponsored by the JREF with performing magicians and a "skeptical" theme. That intrigued me, since I've never believed in gods, psychics, conspiracy theories, dowsing, or homeopathy, and the skeptically-flavored booze on the cruise turned out to be mighty fine. I enjoyed the people I met, made many friends, and was eventually interviewed on several skeptical podcasts exposing some lies told, surprisingly, not *by* but *about* the Discovery Institute.

But getting more to the point, what is scientific skepticism? At its root, skepticism means applying evidence-based reasoning to claims of the supernatural, conspiracy theories, alternative medicine, political assertions, and to any other falsifiable propositions. It means being aware of your own biases; thinking and arguing logically; remaining open to the possibility that you've reached the wrong conclusion; avoiding logical fallacies; and revising your views when the evidence demands. But most importantly it means being

37

skeptical of your own skepticism and of those whose intelligence you admire, because that's when you're most likely to be led astray. As Richard Feynman once said, "The first principle is that you must not fool yourself and you are the easiest person to fool."

Another question that frequently comes up in skeptic circles is whether skepticism necessarily leads to atheism. The answer to this query should be obvious. While evidence-based thinking may lead one to atheism, deism, agnosticism, or ignosticism, I can't think of a single skeptic who applies skepticism to every aspect of his or her life. We are all subject to confirmation bias (along with many other equally insidious biases), and the worst thing is that we're often not even aware of it. Love, political ideologies, religion, and various dogmatic beliefs serve to fulfill emotional needs, and unless such beliefs are falsifiable, they do not fall within the purview of scientific skepticism. Hence it is possible to be a skeptic and believe in god. It is also possible to be a skeptic and be mistaken about various claims. In fact, it would be a cause of great concern if all skeptics thought exactly alike. It is through independent thinking, reasoned disagreement, and civil debate that we are most likely to find the truth, at least to the extent that the truth is accessible to us at this point in time. Needless to say (but I'll say it anyway), there is still much for us to discover and learn, and science appears to be the light leading us along the path to knowledge.

This is why skeptics have adopted a short-form version of the scientific method as a means of testing truth claims. Of course, outside of the laboratory or university setting, without peer review, and without a formal education in science, skeptics cannot apply the scientific method as rigorously as real scientists, but they can still engage in critical thinking and evaluate the available evidence. This may not be a perfect system, but it is the best tool we currently have. Moreover, unlike scientists, skeptics are not subject to the politics of seeking grants or the pressure from private companies, which may have financial interests in seeing particular outcomes from scientific experiments

In conclusion, everyone is a skeptic to some degree. There's nothing special about those of us claiming to be such. Skepticism—even scientific skepticism—is not an exclusive club. At the end of the day, skepticism is just a reminder to think critically and independently, reach decisions based on the evidence, and recognize our own fallibility.

Why You Can't Trust Your Brain
Caleb W. Lack

The human brain is immensely complex, with an estimated 86 billion neurons and some 100 trillion synapses. It is amazingly resilient in the face of injury, able to reroute functions around damaged or purposefully removed areas to restore partial or full functioning. It can store an almost limitless amount of information for decades when healthy. However, despite these wondrous properties, the brain is quite easily fooled. This means that you, yes you, need to seriously doubt... you.

A few years ago, when several students and I started a skeptical/freethought campus group, the Center for Inquiry sent us a giant box of goodies. In it were materials to help us get started: flyers, copies of *Skeptical Inquirer* and *Free Inquiry*, stickers, and more. One of the stickers was a play on the Tide detergent logo and slogan that said "DOUBT—For even your strongest beliefs!" I loved it (and, in full disclosure, kept that one and put it in my office). Doubting is, to me, one of the hallmarks of an enlightened mind and the key to being a good scientist and a good skeptic. It is not enough to only doubt others, which we all do. Instead, people must also be willing and able to doubt their own beliefs and convictions across all areas in life. Many people do not do this, though, and instead plow through their lives convinced that their beliefs and perceptions of the world are accurate and unimpaired. The purpose of this chapter is to teach you some of the many, many reasons why you cannot blindly trust your brain; and to prove that doubt (even of yourself) is a good thing.

Many researchers in the psychological and other sciences have spent decades looking into specific ways that the brain

41

can be fooled. Given the constraints of this chapter, I will not be able to cover all of the identified ways and will instead focus on a number of the more common cognitive biases and mental heuristics that impact us, every day, as humans. At the end of the chapter, I will briefly discuss some methods to help attenuate the impact of these on day-to-day decision making.

What are Cognitive Biases and Mental Heuristics?

Two of the largest factors influencing why you should doubt yourself frequently and thoroughly are cognitive biases and mental heuristics. Since Kahneman and Tversky's (1972) landmark article on how we, as humans, make non-rational decisions, the research on these factors has grown immensely. Cognitive biases are predictable patterns of judgmental deviation that occur in specific situations, which can cause inaccurate interpretation or perception of information. They impact our ability to make accurate, logical, evidence-based decisions on a consistent basis. Recognizing cognitive biases helps us to realize that our intuitive understanding of the world is often (but not always) distorted in some way. Once you understand that you cannot always trust your own judgment, and therefore need to apply doubt to yourself, you are on your way to beginning to overcome some of them. Luckily for us, these are "predictably irrational" (Ariely, 2008) problems, and so, by becoming aware of them on a conscious level, we are able to combat their influence.

Slightly different, heuristics are mental shortcuts or rules of thumb that significantly decrease the mental effort required to solve problems or make decisions (Kahneman,

Slovic, and Tversky, 1982). Unfortunately, this will often lead to an oversimplification of reality that can cause us to make systematic errors that can become cognitive biases.

Although there is not a "standard list" of cognitive biases and heuristics (and in fact there are quite literally hundreds that have been identified), the ones described below are some of the most well-researched and common biases/heuristics we encounter as humans.

- ❖ Confirmation bias
- ❖ Belief perseverance
- ❖ Hindsight bias
- ❖ Representativeness heuristic
- ❖ Availability heuristic
- ❖ Anchoring and adjustment heuristics

Confirmation Bias

The confirmation bias is one of the most encountered, most frustrating, and yet most understandable biases (Nickerson, 1989). It is the tendency of individuals to favor information that confirms their beliefs or ideas and discount that which does not. This means that, when confronted with new information, we tend to do one of two things:

1. If this information confirms what we already believe, we throw it a huge "Welcome home!" party. We unreservedly accept it, and are happy to have been shown it. Even if it has some problems, we forgive and forget those and welcome this new information into our brains with great fanfare. We are also more likely to recall this information later, to help buttress our belief.

2. If this information contradicts what we already believe, we slam the door in its face and tell it to get the hell off of our porch and never come back or we will call the police. We nitpick any possible flaw in the information, even though the same flaw would not get a mention if the information confirmed our beliefs. It also fades quickly from our mind, so that in the future we cannot even recall being exposed to it.

Here's an example. Let's say that you, for whatever reason, think that pit bulls are dangerous animals. When a friend tells you about an accident in which a dog attacked a child, and mentions that it was a pit bull, your brain will glom onto that, and you will say to yourself (and maybe your friend), "See, I knew they were dangerous, terrible beasts!" However, if the same friend later tells you she is dating someone who owns a pit bull, and how well-behaved, sweet, and affectionate it is, you are likely to discount that information in some way ("Oh, she's just twitterpated and isn't objective" or "That dog is an exception to the rule").

The confirmation bias could even be seen as the entire reason that the formal scientific method had to be developed. We naturally try to find support for and prove our beliefs, which can in turn lead to the wholesale discounting or ignoring of contradictory evidence. Science, in contrast, actively tries to *disprove* ideas. The scientific method allows for increased confidence in our findings and makes scientists less prone to the confirmation bias.

If you have ever been in an argument about an issue that you care deeply about, chances are you have experienced an interesting aspect of the confirmation bias: the more emotionally charged or deeply held our beliefs are, the stronger the effect of the confirmation bias (Plous, 1993). This aspect of the confirmation bias underlies one of the primary reasons why reasoning and producing facts does not work very well in most debates and arguments: we have already made up our minds and ignore that information which shows us to be wrong.

Belief Perseverance

Have you ever told someone that something they believed was demonstrably wrong... only to later have them still express their belief in it? Or, perhaps you have clung too tightly to beliefs and ideas even when you should have changed? If so, you are already familiar with the effects of belief perseverance. Stated simply, belief perseverance is the tendency to stick with an initial belief, even after receiving contradictory or disconfirming information about that belief (Anderson, 2007).

Speaking broadly, we as humans tend to show three kinds of belief perseverance: self-impressions, social impressions, and naïve theories. You may believe you have a beautiful singing voice, despite your family evacuating the house when you take a shower (self-impression). Or, perhaps you dislike a co-worker, even after he has done several nice things for you (social impression). Maybe you have a specific belief about how the world works, for example that the Earth is 6,000 or so years old, and even when presented with the overwhelming scientific evidence showing it is closer to 4.5 billion years old you still continue to believe in mythology as truth (naïve theories).

Our beliefs stay believed for a number of reasons, among them the availability heuristic (see below for details on that one), illusory correlations, and distortions of evidence. The illusory correlation refers to seeing a relationship between two things (events, people, places, activities) when in reality there is not one (Eder, Fiedler, and Hamm-Eder, 2011). Superstitions are prime examples: you see a black cat moving across the street, cross its path, and then shortly thereafter have a flat tire, which makes you late for your job. You then look back and think, "I knew I should have gone down a different road!" In reality, of course, unless the cat had ninja-like skills and was able to actually jump onto your car and slice the tire, the cat had nothing to do with it. A more real-world example would be praying for someone you care about to get better in the midst of an illness. Lo and behold, the person does get better, which you then attribute to your intercessory prayer. In reality, carefully controlled studies and meta-analyses have found no link between health outcomes and prayer (Benson et al., 2006; Masters, Spielmans, and Goodson, 2007).

Distortions of evidence can be explained, in part, by the confirmation bias. To continue the above example, because you believe that intercessory prayer is effective, you may remember those times when you prayed and someone got better, but ignore the times your prayers were not or (more often) discount those times as being "beyond my understanding." In short, you remember the hits, forget the misses, and so you distort the actual data in favor of making it support your beliefs.

Hindsight Bias

If you have ever told someone that you knew something was going to happen, after it already did, you have likely fallen prey to the hindsight bias. The hindsight bias occurs when we overestimate how confident we are in an outcome after the outcome is already known (Roese and Vohs, 2012). This bias is commonly described as the "I knew it all along" fallacy.

An example of hindsight bias would be hearing about a couple breaking up and saying, "Oh, I absolutely knew that they were not right all along." Chances are that, before you knew the outcome, you would not have used the term "absolutely" (unless they were just a horrible mismatch). You may have said they "probably" wouldn't make it, or that there "wasn't much of a chance" when they were dating, but once you knew the result, you became much more certain of your previous estimation of the outcome.

Rather than a single construct, there appear to be three distinct forms of hindsight bias: memory distortion, inevitability, and foreseeability (Blank et al., 2008). In memory distortion, we inaccurately recall our earlier estimate of something occurring, which in turn distorts our current estimate ("I said it would happen"). The inevitability form is particularly salient when engaging with many theists, as this bias occurs when one believes a past event was predetermined for some reason ("It had to happen"). The last kind of hindsight bias is the foreseeability form, which shows up when you think that you could have foreseen something that has already occurred as occurring ("I knew it would happen"). Each of us is prone to each of these kinds of bias, but there is interesting work (that I don't have the space to go into here) that shows different types of inputs (or reasons) can impact what type of hindsight bias we show (see Roese and Vhos, 2012 for a sterling review).

Representative Heuristic

At some point in our past, each and every one of us has, despite admonitions against doing so, "judged a book by its cover." You see someone (or something), she/he (or it) appears to belong to a particular category, and so you attribute certain qualities to him/her/it (Kahneman and Tversky, 1972). You are, sometimes, terribly wrong. This heuristic can be thought of as a form of stereotyping, of taking a particularly salient feature of someone and overgeneralizing it inappropriately.

For example, let's say that I, as a professor, see two students in my class. One is well-groomed and dressed, sitting in the front row, appears to be paying close attention to the lecture, and is taking copious notes. The other student is dressed in sweatpants and an old t-shirt, is playing on her phone, and appears oblivious to anything happening in the class. My initial reaction will be to say that the first will do well in class, while the second is likely to do poorly. I am comparing their outward appearances and behaviors to my idea of what a "good" student looks like. In other words, one of them is representative of what I think a good student looks like, and one is not. It could be that the one who seems to be on top of things is completely clueless, while the other is bored because she has read ahead and studied the topic on her own, and wants to get deeper into the issue.

There are several reasons why we misjudge people in this way. First, we ignore the base rates of behaviors, or how common something is in a given population. For instance, I tell you about a person I know who is shy, good at mathematics, and loves Star Trek. I then ask you to guess what his major was in college—business or mechanical engineering? Many people will guess he is an engineer based on the descriptors, and completely ignore that (statistically speaking) he was much more likely to be a business major, as there are many more business than engineering majors on most campuses.

Another reason why the representative heuristic occurs is that we often draw inappropriate conclusions from small sets of data. If I have only met two people from the United Kingdom, and they are similar, I may think that they are representative of all the people from the UK (when they could, in fact, be the only two people like that in the entire British Isles). This process can, of course, be extended to any particular group of people (e.g., race, profession, geography, religion) or objects (e.g., model of vehicle, brand of clothing, style of furniture).

Availability Heuristic

When making decisions, we tend to be biased by information that is easier to recall; such information could be that which is more vivid, well-publicized, or recent. Such easily retrievable information can cause the availability heuristic to occur, when we make judgments about how likely something is to occur based only on how easily it is brought to mind (Schwarz et al., 1991). This can cause a number of problems as to how we process information and make choices.

Much of the research on the availability heuristic has used a protocol similar to the following (adapted from Combs and Slovic, 1979). First, subjects are presented with a question ("Which of the following causes more deaths in the U.S. per year?"). They then have to make a choice between two options in response to this question, one of which is much more easily recalled due to news coverage or recency to an individual (e.g., lung cancer vs. automobile accidents). Even though lung cancer kills three or more times as many people as car crashes each year, a majority of people will tend to choose car crashes as being more prevalent. This choice could be because they can easily recall a crash that they witnessed, were a part of, or saw on television or the news, but have a harder time retrieving information on a lung cancer death.

For many readers of this book, the impact of recency may be particularly salient in activating the availability heuristic. Imagine that you are engaged in a debate, online or in real life, with someone who is diametrically opposed to your viewpoint on religion. The person may be rude, antagonistic, mean, or just unpleasant in some way. You leave the conversation frustrated, and almost immediately encounter someone else who reveals, at the start of your interaction, that he or she is of the same religious view as your prior nemesis. You are very likely to attribute to the new person the characteristics of the old—that they are rude, mean, and so on—without any evidence to back that up.

Anchoring and Adjustment Heuristics

When we try to estimate various phenomena (population of cities, number of murders committed per year, percentage of college students who graduate in four years) that we are uncertain about, our answers can be quite easily manipulated (Epley and Gilovich, 2006). For instance, if I ask you, "Is the percentage of bald Americans more or less than 15%?" instead of, "What is the percentage of bald Americans?" I am likely to get very different answers. This problem is called the anchoring and adjustment heuristic, and demonstrates how our estimates are influenced by initial anchors. It can occur even when given anchors are obviously ridiculous (Strack and Mussweiller, 2006), as in the below example.

Suppose I were to ask you, "Is the percentage of Americans who are atheists greater or less than 0.0005%?" and then I later ask you to give me an estimate of the number of Americans who are atheistic. I then ask your best friend, "Is the percentage of Americans who are atheists greater or less than 95%?" and later ask him or her to give me an estimate of the number of Americans who are atheistic. Odds are, you will give me a much lower number than he will, even though you may have given similar estimates if I had just asked, "What percentage of Americans are atheists?"

Why does this happen? It appears that our estimates stay too close to the original anchor, regardless of its absurdity, and we fail to take into account other sources of information that may help to provide us a more reasonable answer. Our final estimates, then, are off by more than they would be *sans* anchor since we stop once we reach the edge of a plausible range for the estimate. More than likely, the true estimate is closer to the middle of the plausible range but, thanks to the anchor, you stay closer to the side of your anchor.

Conclusion

Although there is not a way to completely rid yourself of cognitive biases, there are a number of tools and methods you can use to mitigate their effects on your everyday decision making. First, doing what you have just done (learning about the biases) is called awareness-raising. It is certainly useful in assisting you in recognizing these biases when they manifest in others' decisions, but unfortunately does little to prevent you from falling prey to them. This appears to be primarily due to the fact that these are largely automatic processes that manifest across almost all aspects of our cognition (Gilovich, 1991; Schatcher, 1999). Indeed, some evolutionary theorists posit that these are not actually cognitive flaws, since many of these processes save time and effort and do not always produce poor decisions and judgments. It is argued that what I have labeled as biases in this chapter have been selected for during our evolutionary history and may be better regarded as resource-conserving, utilitarian, mental shortcuts (Haselton et al., 2009).

Nonetheless, having awareness can at least lead you to engage in some cognitive self-reflection upon your decisions and automatic impulses, causing you to learn to doubt even your own mind and avoid some of the above biases and heuristics (Toplak, West, and Stanovich, 2010). Many people fail to doubt, instead falling into what is called the bias blindspot, recognizing biases in others but reporting that they are less prone to them. Self-reflection alone is not enough, though, as research has also shown that those who are free of the bias blindspot are not automatically unshackled from their own cognitive biases (West, Meserve, and Stanovich, 2012).

If awareness and reflection are not enough, what else can you do? Psychologists have recognized specific methods of decision making that may help to mitigate the effects of our mental shortcuts. The most prominent include training in critical thinking skills, which often emphasize honing one's ability to evaluate evidence independently of one's previously held opinions or beliefs (Sternberg, 2001). Relying on empirical evidence, rather than intuition and "snap" judgments, is another way to avoid potential biases. Allowing empiricism to guide, to the best degree possible, beliefs and decisions is useful because the methods of science are specifically designed to minimize potential biases in hypothesis testing (unlike the human brain).

In conclusion, you (and everyone you encounter in life) will always have cognitive biases and heuristics as part of your mental toolbox. They will often be useful, but can also lead you to make systematic mistakes in decision-making and judgment. Therefore, you can and should try to battle them. Learning to doubt your brain is step number one in this process, but is only the first step in becoming a more effective and less biased decision maker. Training yourself to think critically and employing the use of solid problem solving steps like the scientific method can then allow you to be less prone to the biases and heuristics reviewed above.

Bibliography

Anderson, C.A. (2007). "Belief perseverance." In R. F. Baumeister and K. D. Vohs (Eds.), *Encyclopedia of Social Psychology*, pp. 109-110. Thousand Oaks, CA: Sage.

Ariely, D. (2009). *Predictably irrational: The hidden forces that shape our decisions* (1st ed.). New York: HarperCollins

Benson, H., Dusek, J.A., Sherwood, J.B., Lam, P., Bethea, C.F., et al. (2006). "Study of the Therapeutic Effects of Intercessory Prayer (STEP) in cardiac bypass patients: a multicenter randomized trial of uncertainty and certainty of receiving intercessory prayer." *American Heart Journal, 151*(4), 934-942.

Blank, H., Nestler, S., von Collani, G., and Fischer, V. (2008). "How many hindsight biases are there?" *Cognition, 106,* 1408–1440.

Combs, B., and Slovic, P. (1979). "Newspaper coverage of causes of death." *Journalism Quarterly, 56,* 837-849.

Eder, A.B., Fielder, K., and Hamm-Eder, S. (2011). "Illusory correlations revisited: The role of pseudocontingencies and working-memory capacity." *The Quarterly Journal of Experimental Psychology, 64* (3), 517-532.

Epley, N., and Gilovich, T. (2006). "The Anchoring-and-Adjustment heuristic: Why the adjustments are insufficient." *Psychological Science, 17,* 311-318.

Gilovich, T. (1991). *How We Know What Isn't So: The Fallibility of Human Reason in Everyday Life*. New York, NY: The Free Press.

Hamilton, D. L., and Gifford, R. K. (1976). "Illusory correlation in interpersonal perception: A cognitive basis of stereotypic judgments." *Journal of Experimental Social Psychology, 12,* 392-407.

Haselton, M. G., Bryant, G. A., Wilke, A., Frederick, D. A., Galperin, A., et al. (2009). "Adaptive rationality: An evolutionary perspective on cognitive bias." *Social Cognition 27*(5), 733–763.

Kahneman, D., Slovic, P., and Tversky, A. (1982). *Judgment under uncertainty: Heuristics and biases (1st ed.).* Cambridge University Press.

Kahneman, D., and Tversky, A. (1972). "Subjective probability: A judgment of representativeness." *Cognitive Psychology, 3* (3), 430–454.

Masters, K.S., Spielmans, G.I., and Goodson, J.T. (2006). "Are there demonstrable effects of distant intercessory prayer? A meta-analytic review." *Annals of Behavioral Medicine, 32*(1), 21-26.

Nickerson, R.S. (1998). "Confirmation Bias; A Ubiquitous Phenomenon in Many Guises." *Review of General Psychology, 2* (2): 175–220.

Plous, S. (1993). *The Psychology of Judgment and Decision Making.* Columbus, OH: McGraw-Hill.

Roese, N.J., and Vohs, K.D. (2012). "Hindsight bias." *Perspectives on Psychological Science, 7,* 411-426.

Schwarz, N., Bless, H., Strack, F., Klumpp, G., Rittenauer-Schatka, H., and Simons, A. (1991). "Ease of retrieval as information: Another look at the availability heuristic." *Journal of Personality and Social Psychology, 61,* 195-202.

Strack, F., and Mussweiler, T. (1997). "Explaining the enigmatic anchoring effect: Mechanisms of selective accessibility." *Journal of Personality and Social Psychology*, 73(3), 437-446.

Schacter, D. L. (1999). "The Seven Sins of Memory: Insights From psychology and cognitive neuroscience." *American Psychologist, 54*(3), 182–203.

Sternberg, R. J. (2001). "Why schools should teach for wisdom: The balance theory of wisdom in educational settings." *Educational Psychologist, 36,* 227–245

Toplak, M. E., West, R. F., and Stanovich, K. E. (2011). "The Cognitive Reflection Test as a predictor of performance on heuristics and biases tasks." *Memory and Cognition, 39,* 1275-1289.

West, R. F., Meserve, R. J., and Stanovich, K. E. (2012). "Cognitive sophistication does not attenuate the bias blind spot." *Journal of Personality*

Being Suspicious of Ourselves: Groupthink's Threat to Skepticism
Jacques Rousseau

The skeptical (or freethought, atheist, humanist and so forth) movement has always faced challenges, ranging from relatively trivial issues like the lack of agreement on what those various words mean, to more significant ones such as discrimination against the non-religious.

As various loosely aligned skeptical movements have grown in size and influence, opportunities for internal dissent around strategy or leadership have also increased. So too has opportunity for discrimination internal to the movement (or, opportunity to recognize discrimination that has always been present).

In a world of instant and intercontinental communication via the Internet, dissent can be amplified through the unlimited audience it now enjoys. Simultaneously, opportunities to defuse what might sometimes be simple personality conflicts or misunderstandings are limited by the fact that many of us know each other as avatars or personalities, rather than as people.

While this growth and increased reach is surely a good thing on aggregate, it has also allowed for an increased focus on figuring out where we differ, rather than the focus on what we presumably agree on—the clear goal or goals of promoting scientific skepticism, atheism or humanism.

This is an important moment in the development of the movement, and one that is arguably overdue. The importance is due to the fact that it is not necessarily true that the enemy of my enemy is my friend. So, it is not necessarily advisable to embrace a big tent model where those who belittle and insult religious folk for pleasure are always considered the ally of the thoughtful critic. Nor is it necessarily the case that atheism or skepticism be treated as a good surpassing all others—some of us might want to say that social justice is more (or equally) important, and that those not committed to social justice are no allies at all.

In other words, one sort of threat and one form of groupthink can consist—and has historically consisted— partly of the idea that "we" are a group at all. But as the skeptical movement grows and discovers these differences of opinion, the threat of groupthink does not disappear.

The sorts of groupthink available to us just become more localized, depending on what sorts of allegiances we have discovered. And now that our differences are (at least, to some extent) political, we invest energy and thought into defending our conception of what is important to skepticism, and to defeating opposing conceptions.

In what follows, I describe some ideas that seem to be allowing for a sort of groupthink that encourages disagreement where none need exist. Examples include some usages of the idea of "privilege" to forestall certain forms of criticism, or the accusation—as well as the act—of Internet "trolling[1]" to limit the need for self-reflection and criticism.

My intention is certainly not to argue that these concepts and others are always misapplied, but rather to make the case that insofar as they can be misapplied, we should avoid doing so. There are ways to disagree that do not involve misrepresentation or the self-serving use of words and

---

[1] "Trolling" on the Internet commonly refers to a participant in online debate who is interested more in provocation than in honest and productive debate. While it is not typical to describe someone as a troll if they are simply stubborn in presenting an agenda or cause (rather than explicitly aiming to provoke), I do use the word in this more general and atypical sense in later portions of this text.

concepts in order to buttress ideology—and if we care for reason rather than only rhetoric, any form of groupthink is a threat worth paying attention to.

Convenient Fictions

Even though unreason of various forms can be a dangerous thing, there are few of us who don't occasionally take comfort in some sort of convenient fiction, whether we know it is a fiction or not. It could be the belief that our sports team is the best, or that that the look we received was innocent, rather than evidence of disdain or suspicion.

There is a range of significance to fictions, of course. If you don't much care about what people think of you, you might simply note that look, and think little more on it. You might engage in some casual banter with a supporter of another football team, but not be the sort to have heated arguments about something as inconsequential as sports are.

Fictions that allow you to sweep child molestation under the rug, to justify misogyny, or cause you to pray over a child while she dies instead of rushing her to hospital are clearly deeply significant. In addition to this spectrum of significance, it is to my mind indisputable that whether something is true or not matters. It matters profoundly, because the more false things we believe, the more likely it is that we will make mistakes of various kinds, ranging from the trivial to the profound.

Despite the fact that organized religion is premised on mistakes of various sorts, any attempt to understand religious belief—or the motivations for those beliefs—needs to take into account the value they have for the religious.

But when attempts at interpretation or understanding are obscured by antagonism, stereotypes and caricatures, it is easy to forget that there are more points of agreement than points of difference between us and the average religious person. The average religious person lives a similar life to the non-religious person—caring about their family and friends, trying to be competent or excellent at what they do, whether it is a job or a hobby.

Much of the time, we are aiming for similar outcomes—less poverty, more justice, less sexism and racism, more happiness. Of course there are exceptions, and some are notable—the institutionalized "war on women's bodies" is one, where holding life to be sacred results in opposition to abortion for many Christians. The widespread prohibition of assisted dying is another, and in both these cases the religious view can conflict with the nonreligious one. The point remains that the average religious person is less like the extremist Mullah, or the child-molester-enabling Cardinal, than she is like you and me.

So why is it, then, that the public perception of atheists is that of them being overwhelmingly anti-religious? Part of the answer is that atheism—in and of itself—does not need to concern itself with any of these other goals. It is precisely (and only) about a lack of belief in gods. So it is perhaps unsurprising that what people associate with atheism is merely the proposition that gods don't exist.

And, as atheists, we pride ourselves on not believing in this highly implausible proposition (that gods exist). This tends to mean, at the very least, that we share some minimal commitment to reason in that we want to be guided by the evidence rather than superstition or dogma. And if that is the case, it doesn't seem much of a stretch to suggest that we should apply the same critical mindset to propositions beyond merely the god hypothesis.

So, when we speak of social justice, equality, freedom of speech and so forth, it is perhaps reasonable to expect some similarity in approach, even if not in conclusions reached. To put it plainly, an approach in which we listen to the evidence, and to each other, without pre-judging what someone is going to say, what they believe, or what ideological faction they belong to. In a debate aimed at establishing truth, our opponent's arguments should be assessed on their merits, rather than via indirect and unreliable markers like which websites they frequently comment on, or who they are friends with.

Beware the trolls—and beware mistaking people for trolls

It would be naïve to ignore the fact that some participants in these conversations are not honest brokers. Some are simply unreconstructed trolls, picking at the scabs of others for entertainment. Others are perhaps trolls of a slyer sort, mimicking critical reflection while subtly distracting—and detracting—from the issues that others are trying to address.

Another set of "others" aren't trolls at all—and it seems to me that the community of skeptical or atheist activists and bloggers sometimes has a difficult time of it in distinguishing between these sorts of contributor to the debate. Identifying the latter group is perhaps especially difficult, because even though skepticism should go hand-in-hand with critical and careful reasoning, a positive correlation between the two is far from guaranteed.

The trend on the Internet generally—at least according to my anecdotes—is toward increasing hyperbole and hysteria, perhaps especially so when we can comment anonymously, with little fear of reputational harm. Those who shout the loudest think that they can win, or end up thinking that they've won once they have drowned out the opposing view. And even though our community might (hopefully) be more rational than any randomly selected group, we're not immune to the same trend.

On emotive issues, this can be particularly worrisome, and is also more likely to happen—simply because the stakes are higher. And here is the thing: I think we forget that a concern for tone does not automatically mean that you are a tone-troll[2], and yet, calling someone a tone-troll can work wonders in simply shutting them up—which means that there's less competition for your voice, and your point of view.

To put it another way: you can grant that a young earth creationist (for example) has some pretty confused ideas about which propositions gain epistemic weight via which pieces of evidence, yet still think that it is a bad idea to call him some abusive name. You might think it's a bad idea simply because you think it rude, or you might think that we stand a better chance of persuading an audience to take us seriously if we engage with arguments rather than abuse.

When the space for polite disagreement—even with people who are very wrong—disappears, it is sometimes difficult to imagine the conversation as a rational disagreement. If we hope to change minds (not necessarily the mind of the interlocutor, but of those listening in), we

---

[2] Accusations of being a tone-troll signal that the alleged tone-troll is expressing a discomfort with strong language and (sometimes) insult in Internet discussions as a way to avoid argument, or avoid recognizing the weaknesses in his or her own argument. However, many consider this an open question, in the sense that it is not obviously illegitimate to be concerned with civility in debate.

would be impeding our own progress toward that goal if we stopped listening to each other.

And we're not listening to each other—at least not consistently, or as much as we could do. Right now, the debate on misogyny in the skeptical community has escalated to such an extent that there's a lot that can't be heard over the screaming, and the caricaturing.

To be clear, it is certainly bad if we create, endorse, or fail to combat a climate of hostility to any poorly defined (and heterogeneous, in any case) group like "women." And the fact that some members of our community believe that such a climate currently exists is a problem in itself, whether or not you are complicit in creating that climate.

In fact, it is indeed a problem whether or not such hostility even exists. Unless you want to claim that all instances cited as evidence are complete fabrications, the perception most likely finds inspiration in some forms of behavior or speech that could be modified at little or no cost, even if only through public statements of support, and public condemnations of discriminatory or abusive behavior.

But people do also make honest mistakes, causing unintended offence through not being aware of the particular sensitivities of an author or blog community. And it can be difficult to recognize and apologies for a mistake if you feel besieged. The self-regulating role of blog communities is worth noting: if inappropriate piling in to the abuse of a commenter is not called out, we quickly become gangs who have chosen a side and chosen our authorities or leaders, and who then defend our turf by whatever means necessary—whether principled or not.

This tribalism, and defending of a cause, comes naturally to most of us. What also comes naturally is to double-down when challenged, especially when others question your integrity or motives. This complicates the reactions that people have to being called out for language that appears—or is—sexist or insensitive to the pervasive misogyny debate. Being defensive when accused of sexism is to be expected, and it is perhaps uncharitable to use this defensiveness as further evidence of the commenter's ignorance, prejudice or malice.

Privilege and Its Role in Avoiding Argument

One of the chapters in Bertrand Russell's *Unpopular Essays* (1950) is "The Superior Virtue of the Oppressed." In the essay, Russell[3] criticizes the tendency of those who marched with him in support of various social justice issues to not simply stand against oppression, but also to insist that the oppressed are somehow epistemically privileged. They were wiser, more experienced, and perhaps even more objective than those who were not oppressed. An uncharitable reading (Russell's) would be that it is actually good for you to be oppressed.

We are doing much the same thing if we glamorize folkways, cultures or traditions. And when it is an anthropologist or social critic referring to the "other" in these terms, many of us might be quick to criticize, perhaps asserting that the observer is being patronizing in her glib summaries, born out of a large knowledge deficit.

But what I think we often miss is that the same mistake can manifest in the opposite sort of way. It manifests when we dismiss somebody else's opinion because of their perceived privilege—because they haven't been there, or

---

[3] In addition to being one of the most prominent and influential philosophers to date, Russell was also dedicated to activism in favor of various social and political issues, including nuclear disarmament, anti-Imperialism, and pacifism (the latter of which resulted in a jail sentence during the First World War).

experienced that (namely, the places and things you have, or those you presume to speak for have). Mostly, we make this mistake when we talk about racism and sexism.

To be clear, it is not a mistake to think that in a patriarchal society, a woman is more likely to understand oppression than a man is. It is also not a mistake to think that of a black person in a society like mine, where the economic classes are strongly correlated with race. But it is a mistake to dismiss somebody's opinion on oppression—even your oppression—if they're not black and/or a woman.

In other words, understanding oppression might well be more likely if you are from an oppressed group, but that is not the only route to understanding—and nor does it guarantee understanding. After all, why trust the view from the oppressed perspective to be more reliable than the non-oppressed view? Perhaps oppression brings with it such epistemic distortion that you are less able to understand even your own situation, never mind that of others.

Despite these concerns, one particular sort of dismissal has become commonplace in arguments around oppression, whether on the grounds of race, sex, disability, class or some other version of identity politics. The dismissal works like this: you attempt to shut somebody up by means of a phrase such as "check your privilege," which is meant to simultaneously destabilize their epistemic foundations, while also shaming them. Either or both of these effects result in a rhetorical victory for you, while making it less likely that he will ever again dare to express the heresy in question.

In arguments that pivot on oppression via sexual identity, this trick is accomplished through using a word like "mansplaining" in place of "explaining." If the man then accuses you of being "blinded" by your rage, you could then deploy the word "ableism," which accuses him of thinking blindness to be something negative. If you're really lucky, the man in question would be both white and wealthy, in which case his "neoliberal whiteliness" will hopefully shut him up for good.

Please don't mistake this for a refusal to accept that all of these things can be problems. They can be, and they in fact usually are. There is nevertheless a vast difference between being blind to privilege of various forms on the one hand, and thinking that privilege makes you wrong (or rather, that absence of privilege makes you right) on the other.

Yes, people have different viewpoints, and those viewpoints are always a factor of their class, race, gender and so forth. But if we think it offensive that negative traits are attributed to people because of these secondary characteristics (insert any familiar gender or racial stereotype here), why is it not also wrong to attribute positive traits on the grounds of those characteristics?

You cannot be guaranteed to understand a situation better than someone else simply because you think you inhabit that situation. Yes, it is a factor, and it is a factor that might even contribute to understanding, on average. But it might sometimes blind you to reality through confirmation bias, or through an overly emotive and maybe irrational interpretation. In the meanwhile, someone speaking from a different position might have done sufficient homework, or be sufficiently sensitive, to have a better understanding than you do—even if she is not a "representative" of the group in question.

In other words, we need to separate the issue of epistemic privilege—where nobody is guaranteed to have any, regardless of your identity—from the issue of politics, and the dangers of things like assuming superiority, or offending others who speak from a set of experiences that you have little or no access to. The latter issue is where things like "mansplaining" present a legitimate problem.

But it is also a legitimate problem, and an evasion of your epistemic responsibilities, to refuse to question your own opinions simply because the questions are being raised by a rich, white, heterosexual man (for example). To do so is to take a few external (and often arbitrary) signs as representing the totality of a person and the justification they have for their opinions. It is a bold claim to make that "whiteness" or "maleness" overrides everything else about a person. In fact, in another world or time we might have called these claims racist or sexist.

It is instructive that we never make the claim that these characteristics are relevant when the person who possesses the characteristics happens to agree with us. For example, every time a man expresses support for the idea of a "woman's right to choose" (whether or not to carry a fetus to term), or every time a woman expresses opposition to that same idea, it should count against the simple-minded dictum that men "don't have the right to an opinion" on the topic. The issue is—or should be—good and bad opinions rather than who holds or expresses those opinions.

False Divisions and False Choices

Whether as authors or as readers, we all have to make distinctions between well-meaning interlocutors and trolls, and we all have an incentive to keep websites and blogs free of trollish pestilence. So patience cannot be infinite. Neither, of course, should we make judgments before engaging sincerely and fully with any relevant argument presented to us.

If all we want is to feel self-righteous, and right, it is certainly expedient to make hasty judgments. It is also good to know who the enemy is, and to agree on who that is, because we strengthen communities and minimize dissent through that agreement. But it is also good to change the enemy's mind, where possible, and it's good to discover that someone you thought to be an enemy is actually simply a confused friend.

Those of you who follow the seemingly-endless squabbles in the secular, skeptical, or atheist community will know that fighting with each other is as much a part of the game as combating religious dogma is. And this is not only because there can be dogmatism and unreason on the non-religious side too—which there certainly can be—but also because everyone is sometimes guilty of being more interested in being right than in making progress.

In any area of contestation, caricatures often win out over trying to find common ground. On the pro-science and secular side (and note the false dichotomy there—as if the religious cannot be pro-science, just like pro-life invites the caricature of "anti-life"), what community there is is partly premised on a caricature of the "other," just like religious folk can easily point to some obnoxious atheist they know and use that person as their baseline for understanding non-believers.

What should concern us about these cartoonish versions of reality is, firstly, the possibility that we are forsaking opportunities to learn things—about each other, about difference, about persuasion; and second, that we are impeding progress toward what could in many instances be common goals.

A significant proportion of secular activism—at least on the Web—currently consists of people mindlessly (or so it appears) sharing photographs of a Hitchens or a Sagan looking thoughtful, and accompanied by an inspirational (or blasphemous) quote. Often, these images will come from Facebook groups such as "I fu**ing love science" (as if saying you love something makes it the case that you do).

But many (is it perhaps most?) of the folk doing the recycling of these images don't have a much clearer grasp on the science than the average religious person. Sure, religious folk can have some gaping holes in their understanding of some aspects of science, such as evolution—but in most areas that actually impact on day-to-day existence, they are not quantifiably less well-equipped than the average atheist is.

What sharing photographs of Sagan does, though, is to create a (false) impression of community through imagining that "the other" is an unscientific, Bronze Age-mythology believing monotheist. That "other" in turn is encouraged to construct a shibboleth of the dogmatic, immoral and cruel New Atheist. And we are all sometimes looking through the eyes of our respective prejudices, rather than engaging with the typical believer or nonbeliever.

We perhaps do exactly the same thing with the factions we create or inhabit that are internal to the skeptic and atheist movement. Even though we have various shared goals, so much of our energy seems to be devoted to schisms—creating and perpetuating the "deep rifts" between one imagined hive mind and another.

As time goes by—and as words like "Tom Johnson," "Elevatorgate," "Bunnygate" (as you know, this list can go on for quite some time) start to lose their meaning to all but the most obsessive—the stereotypes nevertheless linger. And the distrust and antagonisms serve not only as fuel for our religious detractors, but are also counterproductive to our outreach goals. Younger atheists and skeptics, looking for a community, might find it difficult to see what we have in common through the persistent fog of war.

The reinforcing and recycling of prejudices—both pertaining to the average religious person, and internally to the various skeptic divisions—is arguably a rather anti-humanist activity. Unfortunately, when combined with the instant gratification (and levelling of power) of social media and the internet, the short-term gains of piling on with criticism and abuse can seem to outweigh the long-term gains of resolving our differences.

I have been working with skeptics and atheists in South Africa for the last 15 or so years, and given that I teach at a university, many of the people who seek me out to talk about these issues are relatively young. And it is fairly consistently the case that what attracts them to the atheist movement is a fair amount of anger, and a desire to express that anger. They feel lied to or betrayed, and feel like they have wasted much time in service of that lie.

To be honest, they sometimes even put me off, because pomposity and arrogance—especially in your average twenty-year-old—is rarely pretty. But because there is a ready-made community of people out there who will validate the anger, and encourage the blasphemy, that arrogance is planted in some very fertile soil. Some atheists seem to never get past that anger, and that arrogance. What this often means is that they never realize that as comforting as it might be to belong to the community of those who are (self-righteous and) right, it is not contributing much to changing the world.

So What Do We Do?

While presenting a lecture sometime in 2011, I introduced the topic of September 11, 2001 to make a point about conspiracy theorists and how frequently they employ confirmation bias to support their views.

A student asked me which argument it was that convinced me the conspiracy theories are false. I replied that the world is not that simple: there often isn't one knockdown argument against a position—especially a position involving so many complexities and confounding details. Instead, I said, it is a matter of the arguments for one position being weaker than the other, when considered in overview.

But sometimes the situation is of course more uncertain than this, and it seems impossible to choose sides on any particular topic. Yet we often choose sides anyway, despite the fact that we cannot support the view that we have chosen to claim. To put it quite plainly, when last have you heard someone, after having defended a view, and when challenged, say something like "I don't know enough about that issue to have a position on it"?

Too many of us seem to despise doubt or uncertainty, even if that is the position best supported by the evidence we have. We leap to conclusions about motives and political allegiances, or an interlocutor's familiarity with some historical debate, because the choice often seems as stark as being silent or being hyperbolic. The options of reserving judgment or trying to argue for a middle ground are too often ignored.

The general point is that we have an option besides that of dogmatic, uninformed zeal—especially in matters that are fraught with political or emotive tension—and that of being expert enough on the topic in question to expect our view to hold sway.

We can say that while we are not certain, we nevertheless think it is likely that a certain position is wrong (or right). This sort of response is premised on more than a hunch or guesswork, but on less than certainty—and never on simply refusing to question our existing beliefs. This more humble approach to what we believe is a constant reminder that it is possible to change our minds when new evidence comes to light. It also sends a powerful signal to others that there is a purpose to debate, and to trying to change your mind on any given issue.

To treat all our beliefs as equally justified—or to forget that we mostly speak from a position of qualified agnosticism—is unlikely to be good for debate and for the possibility of discovering that we are wrong. And if we care about being right, or rather, care about believing things that are true rather than false, we should not forget that getting to the truth is often possible only through allowing ourselves to be uncertain—or even, on occasion, wrong.

Escaping the Filter-bubble

One of the things that the Internet has been good for is broadening the range of perspectives in any given conversation. Of course certain barriers need to be overcome: to participate, you need an Internet connection and a suitable gadget. Nevertheless, conversations have been democratized, thanks at least in part to being able to more easily discover who is interested in talking about the same things as you, and the fact that it is relatively inexpensive to join in.

However, the filter-bubble remains a problem. Not only do the personalization features of search engines like Google give us results that reinforce existing prejudices; confirmation bias also means that—whether we are always aware of it or not—we like it that way.

Now combine that filter-bubble with what seems to be increasing hostility on the Internet, and it becomes a legitimate concern that some voices might withdraw from the conversation entirely. If a congregation of folks with hair-trigger tempers, and who happen to agree with each other, are gathered into one "room," as it were, the prospects for fruitful debate begin to recede sharply.

This has two consequences: the collection of trolls and angry folk are made more homogenous, and they thus appear stronger; and likewise, the collection of those who consider themselves virtuously opposed to the angry folk is furnished with another example of why they are special, and right—and their homogeneity increases too.

So, one day we might end up with half of the Internet grunting angrily at each other, while the other half recites passages from Plato—unless we find some way to arrest this escalation of hostilities, or unless I'm wrong about the trend (and I hope I am). One immediate suggestion is that we should also consider what each of us could or should do, simply in our capacity as members of the atheist community.

First, I would argue that we sometimes place too little or too much emphasis on history, and not enough on our own conduct. "Too little," in the sense of making too little effort to determine whether we are interpreting someone else correctly. And then "too much," in the sense that we sometimes expect new entrants to a conversation to know minute and technical historical details of that conversation—and then abuse them when they get a detail wrong. There is sometimes too little patience for any kind of induction period, and so-called "newbies" need the thickest skins of all.

So when a debate gets heated, we should try to remember that no matter what has come before, we are constantly at a new decision-point, where we—and only we—are responsible for what we say in response to something we find provocative. Sure, someone else has committed a wrong, and we might justifiably be offended.

In expressing our offence, though, we can sometimes forget not only our manners, but also some of the basic rules of engagement such as the principle of charity. You know, that stuff that skeptics pride themselves in being good at.

To some extent, we seem to have lost focus with regard to those rules of engagement because feeling superior, and feeling ourselves to be right, has taken some of the space that used to be occupied by wanting to be reasonable, justified in our views, and therefore more likely to be right.

Asking questions like "who started it" can of course be diverting. Endlessly creating and debating reasons for internal dissent can generate blog views, popular conference speakers and successful podcasts. But they are not the sorts of questions, nor the sorts of activities, that necessarily help to illuminate the more important questions.

Questions like "How can we end it?" and "When will we get back to making the world a better place?"

Science: A Mechanism for Doubting; a Source of Reliability
Kevin McCarthy

The goal of science is not to prove religion wrong. The goal of science is to learn new things about our universe in an organized way. The fact that, during this process, the claims of religions have doubt cast upon them is incidental.

There are scientists who are atheists who do not believe in any religion. There are also scientists who are religious, some are even pastors, who find no conflict between their religious beliefs and their scientific research. In neither case does the scientist seek to support or refute the claims of religion.

That being said, science often makes religion uncomfortable because it does refute many of the claims of religions. All religions have claims that are refuted by modern knowledge. An examination of a common religious belief will serve as an example.

There have been many treatises written about the scientific impossibilities of the Great Flood of the Bible. This story is interesting in that we have some very specific descriptions of what happened in the Great Flood. These descriptions are impossible to reconcile with the laws of physics. The amount of water required to flood the Earth would be more than three times the volume of all the water that currently exists on the Earth. That volume of water would have rained down in a mere 40 days. In our modern age, a third of an inch of rain per hour is considered "heavy." Yet the Great Flood would have generated roughly nine meters (30 feet) of rain per hour.

Knowledge of biology also refutes many of the claims of the Great Flood story. Any fish not on the Noah's Ark would

75

have been killed. Neither fresh water nor salt water fish could adjust to survive in the brackish (and heavily mud-filled) water that appeared in 40 days. Every coral reef on the planet would have died. Every living thing on the planet would have died. The rate of mutation and speciation required to generate the great diversity of organisms we have today, from the organisms that could survive on Noah's Ark, would far outstrip the maximum mutation rates that even evolutionary biologists themselves believe are possible.

Geologists have examined core samples from thousands of locations all over the planet. There is simply no point at which the entire Earth was flooded. Local floods are common and easily seen in the geological record. There are even times when all the land that we see today was underwater. Over 165 million years ago, the Great Plains of the United States, including all of Texas, Colorado, Iowa, the Dakotas, Utah, and Montana and much of central Canada, were covered in the Western Interior Seaway. This was an ocean, in places over 2500 feet deep. The record of this ocean is easily seen in the geological column. Yet a flood that covered the entire Earth to a depth of several miles is entirely missing from that same column.

A final point about the Great Flood is that several other cultures existed at the same time the flood was to have occurred. These cultures both survived the flood and didn't think it worthy of mention in their records.

I choose a Judeo-Christian claim because I am most familiar with them. There are many claims like this in many religions. Claims that are physically impossible. Claims that are recorded once, by fallible human observers (or people with a specific goal in mind), and never seen again. Claims that just do not match processes that we know are in effect today

Some religious people will say that stories like this are metaphorical and are just stories. The problem is that how do we know what, in any holy book, is metaphorical and what is historical. We cannot. These stories and holy books are all open to some form of interpretation.

Some religious people (and some scientists) will claim that these are all "miracles"; that these types of events are products of supernatural powers that are beyond the realm of science. The problem with this is that any supernatural power, even if not directly observable by scientific methods, that affects the material world will leave observable changes. Electrons are too small to be seen by humans, yet we can control and create them with ease. We can see the effects of electrons, for example, in a CRT TV screen. Wind cannot be seen either, yet no one who has lived through a hurricane or tornado doubts that the unseen wind can have massive effects on the world.

But science can, to some degree, test the supernatural. Even if a supernatural entity is not directly testable, the effect of miracles and supernatural activity will include empirical phenomena. In other words, the results of supernatural causes are natural, physical changes. Though the cause may be invisible, the effect most certainly is not (see Fishman 2009 for a more detailed treatment).

In the same way that the invisible wind can cause widespread destruction, any deity who causes miracles will leave traces of its existence. It could have been a miracle that the Flood waters appeared, rained for 40 days and nights, and then dried up. However, there are still no traces for that amount of rain. That amount of rain will leave physical evidence in the geological column. The cultures that have been continuously in existence for the last 10,000 years would have been destroyed. All species on the Earth would show a genetic bottleneck from about the same (very recent) time period. Since these things didn't happen, we have reason to have significant doubts about the reality of a Global Flood.

Scientists, religious or not, do not begin an experiment with preconceived notions of what the results of the experiment will be. Scientists do their best to design and perform experiments that answer a single question. The experimental design removes or controls as many variables as possible, so that there is one thing that changes and there is one result caused by that change.

For example, a company that makes nutritional supplements does not test three new supplements at the same time. There is no way to distinguish the results of one supplement from the others. They would not even test one supplement, but instead one ingredient of the supplement at a time. In this way, they can verify whether the ingredient is responsible for the effect. This is called "correlation." A high correlation means that it is very likely that the variable is responsible for the change. An experiment designed in this way is called a "controlled" experiment.

Another factor science acknowledges is that humans are notoriously prone to confirmation bias (see Caleb Lack's excellent essay in this book on the fallibility of the brain); finding patterns where none exist; and are prone to logical fallacies that seem to support their personal beliefs and help them reject other beliefs. The scientific process works best when humans are removed from the experiments or observations. Do not let a human use a stopwatch. There are too many variables that creep into the experiment. Everything from reaction time to tiredness can affect the results. Instead, use an electronic timing device, where the human just flips the switch and walks away. This is the best way to conduct scientific investigations.

Another way to remove human bias is modern research methodology, called the "double-blind study." It is often used in psychological and medical research. This technique means that not only do the patients not know which drug they are getting, but their attending physician does not know either.

All the physician knows is that patient A is getting a shot from vial #1. The double-blind study removes all bias effects, including the placebo effect, from the study. The doctor and patient report how they are feeling and actual medical improvements (or no improvement) without the chance that a bias ("I hope this medicine works because I have just invested in the company") may creep into the study.

When investigated without the biases of humans and under controlled conditions, claims of religions fail. For example, no god, goddess, deity, or alien power has ever been observed to heal people on request. Scientific research on the healing power of prayer has been going on since the late 1800s. In the case of studies on the effect of intercessory prayer (prayer to help someone else) on health, neither the doctor nor the patient know if the patient is being prayed for.

Two studies (Masters 2006 and Hodge 2007) both examined multiple previous studies (called a meta-analysis). These researchers looked at both the results and the methodology for prior studies of intercessory prayer. The 2006 study examined fourteen prior research efforts and found zero statistical evidence that intercessory prayer was at all effective. The 2007 study concluded that the results are inconclusive.

A 2006 study by Harvard professor Herbert Benson may be the most rigorous study on intercessory prayer. Over 1,800 cardiac patients were randomly divided into three groups. Members of the first two groups were told that they might receive intercessory prayers. The third group was told that they would receive the prayers and did receive them. The researchers examined each patients' records for major complications and checked the survival rate for 30 days after surgery. The third group had the highest incident of major complications and deaths within 30 days of surgery.

Many proponents of intercessory prayer would look at the inconclusive results as a good sign, but that is much too charitable. There are many cases in the mythology of religions of people being healed from horrible diseases and returned from the dead after several days. Those are not inconclusive results. There is no verifiable record anywhere of someone going into surgery after a massive trauma and walking out of the hospital the next morning, completely healed, due to prayer. There is no modern record of prayer curing AIDS, regrowing an amputated limb, or restoring sight to a blind person.

There are many cases of people praying over someone and their cancer going into remission, but that does not mean that the prayer caused the remission. The biology of cancers, the body's response to cancers, and environmental effects are extremely complex. Cancers sometimes go into remission.

Another aspect of cancer studies is that most of the time, the people being prayed for are also undergoing medical assistance in the form of chemotherapy, radiation treatments, surgery, and pain relief. With all of those additional factors, it is impossible to say that prayer was the single factor that caused the cancer to disappear.

So here is a real life example of how a non-critical approach to prayer can take form. I met a committed Christian at a talk recently who was extolling the virtue of prayer. He claimed that a close friend of his (who was "on the edge of Christianity") had a very serious leg problem (which his father had died from). The friend had come round and had been prayed for by this first man and his wife. Soon after, the man's leg recovered from this very serious problem. And this proved, to the first man, that prayer worked. It also served to bring the friend into the fold of committed faith.

Of course, I am sure that you can see the issues. I pointed out to him that in order to arrive reliably at that conclusion, and for the scientific method to do so, would require (at least) that a large number of similar cases were examined; that one would need to know the natural recovery rates for such issues; that one would need to know the recovery rates for people prayed for in a controlled comparison; that one would need to analyze after what period the subject had been prayed for that prayer could still be validly claimed as being responsible for the healing; that one would need to know what other care was being provided as well, and so on. How do we know, as mentioned earlier, that the prayer, and not medical care, was responsible for the healing? If he was prayed for and healing took place one week, month or year later, what would qualify as successful prayer? The scientific method looks to methodically decipher relationships of cause and effect in this way to arrive at reliable conclusions.

What about cases of healing in which medical assistance was not provided? Fortunately, modern medical research is prevented from conducting studies that put a person's life at risk, for example, by not providing scientifically-supported medical help. However, there are a few people who, for religious reasons, have refused medical help. Unfortunately, this happens to children to an alarming degree as parents refuse medical treatment to their children on religious grounds.

One study (Asser, 1998) examined 172 child fatalities between 1975 and 1995. Of the 172 children who died, 140 would have had a 90% or better survival rate with medical intervention. Eighteen more would have had a 50% or better survival rate, and all but three children would have had at least some benefit from medical care. Instead, every one of them died because their parents chose faith healing over evidence-based medicine.

What evidence would be acceptable of the healing power of prayer or of a deity?

Let's say that there is a religion that says that its deity of choice will heal the sick members of that religion. In that

81

case, then we would expect to see no cancers, no disease, no illness of members of that religion. One or two might be an acceptable rate of illness as there may not be 100% belief in the deity. We would expect to see members of that religion regrow limbs, recover from major trauma in very short amounts of time, and not suffer from flu, colds, or other unpleasant diseases.

If there was a religion that had this kind of healthy membership and verifiable stories of healing, then even science would show that the healing power of that deity existed... or at least something that was going on within that religion. Science could at least test and find an unknown cause, an anomaly, in these situations of empirically observable healings.

People all over the world have made claims of how prayer healed themselves or a loved one. However, the evidence is very clear, that there is no beneficial effect of intercessory prayer. Does this mean that science has disproved the existence of deities? No. But it does mean that one of the claims of many deities (healing of believers) has no evidential support.

An analogy will show how claims have doubt cast upon them by examination.

A friend has ten boxes and says that nine of them have $100 bills in them. You may pick one and keep what is inside. The first box is opened and nothing is in it. Well, you think, I just got that 1 in 10 box. But then you check another box and that one is empty too. Now, you have a firm fact that your friend is not telling the truth. You have serious reason to doubt your friend's claim that nine of the boxes have $100 bills in them.

Then you check all the boxes and one of them has a dime in it. Now, there is no doubt. Your friend's claim has been completely disproved. It is not possible for your friend to be telling the truth at this point. He made a claim and you examined that claim and found it to be not true.

That is what science does. That is its entire purpose—to examine claims and determine whether they are valid. In science, this claim is called a "hypothesis." The purpose of an experiment is not to prove a hypothesis correct, but to prove it wrong. Even theories which have massive amounts of evidential support are not actually proven to be true.

Instead, scientists think that a particular theory is true because it has a great deal of evidential, experimental, and observational support. New information could come to light that would cast doubt on the theory. In practice, this is very rare because of the large amounts of supporting evidence needed to create a theory in the first place.

At a more fundamental level, how do we learn things about the universe? The only method of learning about how the universe works is science. "Methodological naturalism" is the technical term. Literally speaking, this is the method of learning about the physical/natural world. Every tool and invention that exists in human culture exists because of science. Very simply, no other method has been shown to reliably produce the same results every single time.

No one goes out to their car and prays for gasoline to be explosive under pressure. Gasoline vapor, when mixed with oxygen and ignited with a spark, explodes. Every time. That is the way the universe works. It works for every person, in every situation, no matter the person's culture, beliefs, location or physical or mental characteristics.

Many religious people talk about "revelation" as a source of knowledge. The problem with revelation is that different people have received different revelations. This source of knowledge does depend on what your culture is. The revelations received by those who worshiped the ancient Greek gods were very different from those who worshiped the ancient Norse gods.

We have myths of turtles supporting the Earth and horses with the tail of a peacock. We have myths of women turned into spiders and men with the head of a bull. We have myths of how to cure disease and how to anoint the dead.

None of these myths reflect scientific reality. Yes, even the myths of the most popular religions in the world today. Sure, they may have some historical facts correct. But if that were all that is required for everything to be true, then the movie *Titanic* was a historical documentary, not a work of historical fiction. A few sparse facts that are known to be true do not make an entire work become the truth.

What does reflect reality? Science. When science looks at the myths, the claims, and the history of the world's religions, it finds problems. Problems that deal directly with the claims of the religions.

The night sky is not a tent with holes poked in it. The night sky is infinite. Those little points of light are stars; many (perhaps most) have their own planets circling them.

The Earth is not flat and it was not formed in a few days and the universe is not ten thousand years old. We can see that Earth is a sphere; even the ancients knew that to be true. We know how planets form, we know the chemical and physical processes involved, and we can see those processes in place on our Earth. We can actually trace the history of the universe by observing the past. We can really see what the universe was like ten billion years ago.

Each of the claims that I just made has tens of thousands of peer-reviewed papers to support it. Some would claim that saying that a scientist says something is an argument from authority. It is not an argument from authority to cite these papers and the consensus of every expert in the fields of geology, astronomy, and cosmology. In the hundreds of years that humans have been staring into the universe, no one has ever determined that any of the statements I've claimed here are incorrect.

Part of the process of modern science is peer-review. When a scientist makes a claim and provides evidence to support that claim, the data are given to other experts in the same field. Those scientists then review the experiment and look for flaws. They review the conclusions and judge if the conclusions are appropriate. Other scientists then repeat the experiments and hopefully get the same data and come to the same conclusions. When all of this has been done, then the original work is judged as being verified.

Why should we trust science? Haven't there been cases where scientists attempted fraud? Yes, there have been and every single one of those frauds was revealed by other scientists. Science is self-correcting. And I cannot stress the importance of this enough. Science (or more accurately, the scientific method) as an epistemological method has built into its very structure a process of refinement. Over millennia, humans have been gathering data, making hypotheses, disproving or refining them such that our knowledge base becomes ever more accurate. Mistakes or faulty hypotheses are ironed out. A great example of this is the CERN experiment which found that neutrinos travel faster than the speed of light. The experimenters challenged people to disprove the results in some way, or find the fault. Either fundamental physical theories would have to be refined or flaws would need to be exposed in the experiment itself. In the end the velocities of the neutrinos observed were likely due to a faulty connection in an optical fiber of the Master Clock. Self-correcting mechanisms lead to more accurate conclusions.

But religious claims almost always work in complete reverse to this. Religions start with a claim and hold to it dogmatically, in spite of evidence. Biblical literalists will take the word of a two-thousand-year-old book with regard to a global flood (or any other claim), without correcting or refining this conclusion. This, in spite of the evidence to the contrary. There is no self-correcting mechanism built in. And one cannot claim that the evolution of theology over millennia reveals the operation of a self-correcting mechanism. If anything, it shows a pattern of ad hoc reinterpretation in the face of contrary evidence or thought. Think of the pea in the shell game being moved around by the confidence trickster.

What is amazing about science is that nothing but the data matters. Two scientists may not speak the same language, but they can see that the data from their two experiments support each other. Science is the same in China, India, the United States of America, France, and every other country on the planet, planet in the solar system, and galaxy in the universe.

Because of their very nature, religions are not like this. A religious practice that makes one person feel better may not help another person. Two churches, even in the same faith, may promote different interpretations of a passage or an event. Different regions will have differences of opinion. There is no consistency. In fact, many religious practices and claims are mutually exclusive.

When we look at science, we see consistent processes. We see conclusions driven by data. Most importantly, we see results that work time and again in the real world. When we look at religion, we see none of these things.

Science does not make it a mission to disprove religion. That is merely a side-effect of the study of the natural world. Probably, one of the reasons religions began was to explain things that early humans could not understand. Now we have an effective tool to help us understand—the scientific method. And when the vague revelations of religion are compared to the harsh reality of science, there is a reason to doubt.

Bibliography

Asser, S. M., and Swan, R. (1998). "Child fatalities from religion-motivated medical neglect." *Pediatrics,* 101 (4 Pt 1), 625–9, retrieved from http://www.ncbi.nlm.nih.gov/pubmed/9521945.

Benson H, Dusek JA, Sherwood JB, et al. (April 2006). "Study of the Therapeutic Effects of Intercessory Prayer (STEP) in cardiac bypass patients: a multicenter randomized trial of uncertainty and certainty of receiving intercessory prayer." *American Heart Journal,* 151 (4): 934–42. doi:10.1016/j.ahj.2005.05.028.

David R. Hodge, "A Systematic Review of the Empirical Literature on Intercessory Prayer" in *Research on Social Work Practice March,* 2007 vol. 17 no. 2 174-187 doi:10.1177/1049731506296170.

Fishman, Yonatan. (2007) "Can Science Test Supernatural Worldviews?" *Science and Education,* 2007 Jun; 813-837 doi: 10.1007/s11191-007-9108-4.

K. Masters, G. Spielmans, J. Goodson. (2006). "Are there demonstrable effects of distant intercessory prayer? A meta-analytic review." *Annals of Behavioral Medicine,* 2006 Aug; 32(1):21-6.

Science is Predicated on the Non-Magical Natural World Order

John W. Loftus

I'm going to make the case that science is predicated on the non-magical natural world order. I'll begin by speaking about fairies and how science is eliminating invisible magical supernatural beings, or agents, or gods, as explanations for sicknesses. I'll go on to explain why science works so well, why it cannot work with supernatural agent explanations, and show why supernatural agent explanations are pseudoscientific ones.

Fairies and Invisible Magical Beings

My Irish heritage is rich with superstitious fairy folklore, most notably Leprechauns. There is even a *National Leprechaun Museum* in Dublin, Ireland. Fairies were both feared and respected. But then most cultures have believed in them. Ever hear of Santa's elves? Most every civilization has had its own collection of elf and fairy myths.

To become better informed than an idiot about them, read Sirona Knight's book, *The Complete Idiot's Guide to Elves and Fairies* (2005). Knight is a modern New Age witch, who wrote her book so readers could learn about these magical creatures and how to best "encounter" them. "Key topics include fairy magic in the 21st century, how to recognize an elf and what to do when you meet one, how to attract good elves and fairies, and how to protect yourself from bad ones."

Fairies are believed to cause misery. If you offend one of them they may throw a dart at you. Fairy darts are believed to cause severe pain and/or swelling in the hands or feet. It's believed fairies may cause people to have sinus infections, strokes, and even tuberculosis. If someone catches a Leprechaun to learn where he hides his treasure, you will be sorry. They can very nasty, it's believed.

Fairy Paths are alleged to be the routes fairies use to get from place to place. Never mistakenly build a house on one or you will not have any peace in that home. When building a house the best way to avoid a fairy path is to set four posts at the corners of the site overnight. If all four posts are still standing when dawn comes it's safe to build your house there, otherwise pick a different spot. People who have suffered sickness for no apparent reason have decided that their house was mistakenly built on a fairy path. What to do? Board up the front door and build a new one on another side of the house. If you do this then when the fairies come up to the boarded up door they have to go around it.

Supernatural magical beings like the fairies of Ireland, the trolls of Norway, the spirits of Haitian voodoo, and the evil thetans of Scientology are numerous in today's world and they litter the ancient highway. People believe in them just as much as others in today's world believe in their invisible saints, angels, demons and gods. These magical beings are used by believers as explanations for unexplained sicknesses and healings, with no way to determine which one of them, if any of them at all, was involved.

As science began finding cures for what ails us, so too did the belief in fairies and other supernatural magical beings begin to decline. What is there about the advancement of science that does this? One of the main reasons is because science explains the causes for sickness far better than supernatural magical beings do. Science explains these phenomena so well that the supernatural invisible magical agent hypothesis becomes an unnecessary one at best, and, it follows, these beings or agents probably don't exist at all.

Why Does Science Work So Well?

Pierre-Simon Laplace (1749–1827) is remembered as one of the greatest scientists of all time. He's referred to as the French Newton, or the Newton of France. When Napoleon had asked why he hadn't mentioned God in his discourse on the orbits of Saturn and Jupiter, he is quoted as saying: "I had no need of that hypothesis." This often quoted sentence probably best describes why science works so well. It doesn't require, or even seek out, an invisible magical being explanation to "save the phenomena."

This is how science must proceed if it is to work so well. It *shouldn't* depend on any supernatural magical invisible agent explanations. As David Eller argues, "Science is essentially and necessarily a-theistic: it knows no god(s), admits no god(s) and needs no god(s).[1] He explains:

> When religious entities really are 'invisible' (to all our senses), or to the extent that they really are outside the scope of nature, *they are also undetectable.* But what can we possibly do with the undetectable? How can we study it? How can we know about it—or even know if it exists? We cannot. It is a null category as far as human knowledge is concerned. Science, therefore, limits itself to the detectable because that is the only thing we as humans have access to and can know.[2]

---

[1] David Eller, *Atheism Advanced: Further Thoughts of a Freethinker* (Cranford, New Jersey, American Atheist Press, 2007), p. 202.
[2] Ibid., p. 211.

Because of this he says, "You can have any belief you like, but you *cannot introduce any type of explanation* you like into your science and still have it to be science. In other words, you might believe that a god makes hurricanes, but if you explain actual hurricanes with an appeal to god(s) you are no longer doing science."[3] He explains: "The entire project of science depends on the regularity and predictability of nature, and [supernatural] agency makes events irregular and unpredictable."[4] Science, to be science, must be predicated on the non-magical natural world order. It cannot progress any other way. Believers may not like this and may additionally claim faith is a virtue despite science. But this is how science has progressed so far, and it must be how science will progress into the future, if it's to progress at all.

Just think how science could possibly work if it punted to supernatural agents as explanations. These agents would have their own goals, aims, desires, interests or wills. With supernatural agents, as Eller explains, "we never know quite what they will do. The exact same situations can lead to opposite results if the agents so choose; there is no connection between causes and events. This frustrates and precludes the possibility of ever knowing with any degree of confidence what will happen next."[5]

---

[3] Ibid., p. 210.
[4] Ibid., p. 214.
[5] Ibid. As the reader can tell, I highly recommend chapter 6 of Eller's book, titled: "The Atheism at the Heart of Science" pp. 199-235.

Science Denialism

Believers have to be ignorant of, or denigrate, or deny, science somewhere along the line in order to maintain their beliefs. They will say that if science is predicated on the non-magical natural world order then it excludes one possible explanation for sickness. They will say that science cannot exclude any possible answer and still be considered science. If it does then it is unduly prejudiced against actually knowing the true nature of the reality it tries to understand. If magical beings and/or supernatural agents are prejudicially taken off the table *a priori* ("from the first" or before looking) then science is not trying to understand reality at all, that it has anti-supernatural bias with anti-magical blinders on.

However, science merely attempts to explain things like sicknesses. It did not start off in its infancy by excluding magical beings and/or supernatural agents. Many early scientists were believers.[6] It's just that over the centuries scientists have concluded that for science to work at all they must exclude them as explanations. The progress of science has had a cumulative effect on science itself. Sorry about this, but science explains the phenomena better than magical beings or god(s) do.

Believers cannot legitimately complain about this. What they need to do instead, is show how their invisible magical beings provide better explanations for the phenomena under examination. Believers need to show how faith produces knowledge. What is its method? Why is it people of faith disagree and cannot settle their own differences? Why is it

---

[6] No, Christianity was not the cause for the origins of science. Just read chapter 15 by Richard Carrier on the topic in my anthology, *The Christian Delusion: Why Faith Fails* (Amherst, NY: Prometheus Books, 2012), pp. 396-419.

that with faith almost anything can be believed or denied? What progress has faith made over the centuries? Considering these types of things, their complaints are utterly without foundation.

The truth of the matter is that although science cannot test whether invisible magical beings exist, since by definition they are not detectable by empirical methods, science can and does test magical and/or supernatural explanations of phenomena. It tests the efficacy of petitionary prayers. It tests the claims of the paranormal. It does so in the only way it can, by trying to detect the detectable. If a magical being or supernatural agent has acted in the world then that's where science can kick in. Yet when science gets involved by looking for objective evidence that a supernatural agent has acted in our world it doesn't find any. Joe Nickell has spent most of his life testing these claims as the only full-time paranormal investigator. The results can be seen in his massive book, *The Science of Miracles: Investigating the Incredible.*[7] Anecdotal evidence and/or personal private experiences provide no objective evidence for any invisible magical being out of the many that are believed to exist.

If looking for and not finding unicorns, hobbits, the Loch Ness Monster or Big Foot is strong evidence that they don't exist, then looking for and not finding an invisible supernatural agent, or a god, is equally strong evidence that such agents don't exist either.

Science Made Impossible

If there are one or more invisible magical beings who regularly intervene in our world then science would never

---

[7] Published by *Prometheus Books*, 2013.

have originated in the first place. In Indiana where I live, every year we're hit by several thunderstorms in the warmer months. You've seen them approaching. You know their effects. I got to thinking as one approached as predicted by live Doppler Radar, what did a god have to do with it? When did he do what was required in the atmosphere to send it? Did he make it rain here because of anything we had done, or not done? Did he decide not to send the rain elsewhere because of what others had done, or not done? Did he intervene in the natural process of the world order to do so? When did this whole meteorological process start? Wherever it started meteorologists would be baffled. One minute they would observe a calm clear sky and then, out of the blue, would appear the atmospheric conditions leading up to a storm. If a supernatural agent did that very often then the science of meteorology would never have originated. Meteorologists could not predict weather patterns and storms at all.

We know what the ancients believed. They believed that the gods literally sent the rain, bumper crops, and that they opened wombs. Conversely, it was believed the gods sent droughts, famines, and closed women's wombs. Well, do they? If so, they must continually intervene in the natural order of the world. If so, modern science would never have arisen in the first place. Modern meteorology would not be possible. Modern agriculture would not be possible. Modern obstetrics would not be possible.

The fact is that it didn't have to turn out that science works. A god of some kind could have made science impossible by intervening into our daily lives just as ancient superstitious people thought he did. That modern science originated is evidence that an intervening supernatural magical being does not exist.

Modern astronomy arose because Copernicus and Galileo were looking at the planets (i.e., literally "deceivers") and trying to best figure out how they moved. Prior to their discoveries it was thought the planets would back up for a period of time then move forward again in endless retrograde circles in the sky. The Ptolemaic universe made no sense of the phenomena. But if a god made the planets move like this without any natural explanation or mechanism then the god-explanation would stop astronomy dead in its tracks. The truth, as we know, is that the planets merely appear to back up and then move forward again. This is explained because the sun is the center of our solar system. The planets pass each other in their yearly trips around it. When astronomers figured this out modern astronomy was born, and it was predicated on the non-magical natural world order. Those early astronomers were believers but that doesn't matter. They predicated their calculations on a natural explanation for the movement of the planets.

Modern medicine arose predicated on the non-magical natural world order, too. Just imagine this if you will. Let's say a god healed people based on their prayers. If he did that regularly then taking a pill or not taking one would be irrelevant to whether someone gets better. If God had decided someone didn't deserve healing then even if a pill was swallowed it would have no effect on them. So doctors would be baffled as to why some people got better from administering penicillin while others did not. Doctors could not conclude penicillin heals people and modern medicine would never have arisen in the first place. The very fact that modern medicine can detect illnesses and heal bodies is predicated on the non-magical workings of the human body. Likewise, the very fact that neurologists can detect brain malfunctions and correct them is predicated on the non-magical workings of the human brain.

Scientists do not detect these kinds of supernatural interventions in any of their areas of expertise. If a god had done these things as modern science was beginning then modern science would never have arisen at all.

Not only this, but if natural explanations for events were not possible because a god regularly intervened in our world, then to the same degree science would not be possible in today's world either. But science is not only possible, it has amassed an impressive amount of knowledge which has produced our modern world. So how likely is it that a god has intervened compared with the weight of knowledge science has produced? At best, if a god has intervened in our world then he has done so in such minimal ways as to be indistinguishable from him not intervening at all.

Agreed? You should! Otherwise throw away your cell phones, tablets, and laptops. Forget about your oven or microwave or toaster. Forget about trains, planes, and automobiles. Don't bother with rocket science either. The only reason we rely on this technology is because science works. All we need to do is ask why it works and we'll see it is predicated on the non-magical, non-supernatural workings of the natural world. So in the end, I think we must choose between god(s) or science. We cannot have both.

A Creationist's Objections

Undeterred, Dr. Vincent Torley, who writes for William Dembski's blog *Uncommon Descent*, has written a rebuttal to my claim in this chapter.

The unresolved question is how much his god intervenes and the kind of interventions required of his particular god-hypothesis. Torley argues that the quantity and quality of divine interventions required for the universe to produce all creatures great and small do not render creationist science impossible. Oh, but to the contrary, it has. Given the assumption that a god has intervened to create all creatures great and small then creation science would not be considered to be a science. And guess what, it isn't. The overwhelming consensus of scientists is that if a god has intervened the number of times required then creation science is not possible as a science.

Creation science is not science precisely because it's premised on the intervention of a supernatural agent, or a god. It may be something else, and a god may have indeed intervened to create all creatures great and small, but they can't go around saying what they're doing is scientific. It cannot be. In order to maintain their faith, believers like Torley must argue that science doesn't tell the whole story. They must argue there is a different method for knowing the truth about the universe that is not based on science itself. And their god has knowingly placed them in this position since he presumably knew about the rise of modern science when creating the universe. So his god also knew Torley and other creationists would not be considered by scientists to be doing science at all.

Here's the essence of Torley's criticism:

> All Loftus' argument proves is that if God intervenes in the world, He does so relatively infrequently...The point I wanted to make is that even if we postulate 10 million separate interventions in the 4,000 million-year history of life on Earth, that would still work out at only one act of Divine intervention every 400 years. If I were a scientist, I wouldn't be too troubled by that...Why, then, should scientists be perturbed by supernatural events that occur

once every 400 years, especially when these miracles don't affect their laboratory experiments?[8]

The last sentence is interesting. Torley is contrasting the past, which presumably could represent 10 million divine interventions, with present day laboratory experiments that are taking place between these supposed interventions. The question he cannot answer is why we should think a divine tinkerer intervened at all when he doesn't do so now. If there is no evidence he does so now, then there is no reason to suppose he did so in the past either. And if a god had intervened as often as Torley supposes then evolution would never have come to be established as the fact it is. He disputes this but I don't see how. Since evolutionary science is a science then it wouldn't have won over so many scientists if his particular god intervened so many times.

Furthermore, if there have been 10 million divine interventions in the past then not only would evolutionary science be impossible, so would astronomy, geology, paleontology, and plate tectonics, since if a god had intervened in these areas to the same degree those sciences would not be possible. Since these sciences have produced a massive amount of knowledge then a god probably did not intervene in the past in these areas. Torley needs to show why the science of astronomy, paleontology, geology, and plate tectonics detect no divine intervening hand but when it

---

[8] See "Detecting design (2): A reply to John Loftus," at
http://www.uncommondescent.com/intelligent-design/detecting-design-2-a-reply-to-john-loftus/. See also: "Why Creation Science is Pseudoscience With No Ifs Ands or Buts About It," at Skeptic Ink:
http://www.skepticink.com/debunkingchristianity/2012/12/02/why-creation-science-is-pseudoscience-with-no-ifs-ands-or-buts-about-it/#sthash.zCthNGIC.dpuf. See also, "Heads I Win Tails You Lose, Another Christian Apologist's Trick," at
http://debunkingchristianity.blogspot.com/2012/11/heads-i-win-tails-you-lose-another.html

comes to biological evolution there is one. Torley must therefore arbitrarily exempt these sciences from his god's invisible intervening hand. Upon what basis can he do this? Faith is no excuse when doing science. Faith makes science impossible because it is predicated on the non-magical natural world order.

Science is Nonsectarian by Nature

Another way to say what I'm arguing for is that one of the major reasons why science works is because it's nonsectarian by nature. Science cannot be placed into the service of a given religious or paranormal sect if it's to be considered science. Sectarian "science" is pseudoscience. Sectarian "science" is an oxymoron. Once we allow science to become sectarian then any supernatural agent and any religion or paranormal agent will do.

How we can we tell pseudoscience from science? That should be easy. No one but believers within a particular religious or paranormal sect agrees with it. That's why it's sectarian pseudoscience in the first place. Scientology is based on pseudoscience because no one but a Scientologist believes evil body thetans infest our bodies. Jehovah's Witness accept pseudoscience because no one but them believes blood transfusions are unhealthy and unwise. Mormons accept pseudoscience in that no one but them believes the Book of Mormon describes ancient historical events in the Americas, since the archeological evidence says otherwise. The same thing goes for Mormonism when it comes to DNA testing, which proves Native Americans are not descendants of Semitic people, or that magic underwear protects them from harm. These are only a few examples. Many more could be supplied.

What motivates sectarian pseudoscience? It comes from the need for a particular religious or paranormal sect to gain credibility. It stems from their implicit recognition that science produces reliable knowledge. Otherwise, why do they appeal to science in the first place? Why don't these sects just admit faith is the basis for what they believe? Why not proudly proclaim they don't have the support of science? Why not just extol the virtues of faith and totally denigrate or deny science rather than pay lip service to it? By creating sectarian pseudoscience it already gives the game away. Just like counterfeiting money is valuable (until caught) so also religionists counterfeit science.

Religionists acknowledge this when seeing sectarian pseudoscience in other religions and paranormal sects. They just don't see it when it comes to their own particular sectarian pseudoscience. They operate by a hypocritical double standard. Each sect accepts the results of science when examining all religions and paranormal sects *except their own*. They should be consistent by accepting the results of science across the board. That's why I have proposed the *Outsider Test for Faith*, which equally applies to how these sects examine each other's sectarian pseudoscientific claims.[9] Only if they consistently apply the results of science can they know which religion is true, if there is one.

Conclusion

Science is predicated on the non-magical natural world order. It has to be. It cannot do otherwise. Believers cannot reject it unless they can provide a better alternative. Since faith has no method and produces no reliable knowledge, the only intellectually honest thing for believers of all stripes to

---

[9] "The only way to rationally test one's culturally adopted religious faith is from the perspective of an outsider, a nonbeliever, with the same level of reasonable skepticism believers already use when examining the other religious faiths they reject." See John W. Loftus (2013), *The Outsider Test for Faith: How to Know Which Religion is True*, Amherst, NY: Prometheus Books.

do is to bite the bullet. Accept it. Embrace it. Trust it. There is no other alternative for gaining objective knowledge about our world but science, the only kind of method that really matters.

The Power of Hume's *On Miracles*
Zachary Sloss

David Hume's *Of Miracles* is a universal and powerful attack on the strength of the evidence we have for miraculous events, concentrating on the primary source of knowledge of such events: the testimonies of those present at the time.

Agency and the Laws of Nature

In order to fully understand the power of Hume's argument, we are required to grasp just what is meant by a "miracle" and how that definition is utilized by the argument. We should not be content with defining a miracle as simply an apparent "violation of the laws of nature" (p. 111). If we observe an instance where a natural object does not behave according to the physical laws we had hitherto discerned, we would either consider our senses to have been momentarily deceived or rethink our conception of the relevant laws. It would belie what we mean by "law" if we were to assume that it could be violated at random.

What we require, therefore, is a reason to think that such a violation is purposive. To borrow Hume's example (p. 122), if there is a global darkness lasting eight days, we would not suppose that we had witnessed a miracle (Hume suggests that this might be a miraculous event, of a sort that might be justly believed) since it is unclear that the darkness serves any purpose at all. While it may invoke the birth of new cults, the wiser among us would take a cautious approach and investigate the natural cause of the phenomenon. To convince us that natural law has genuinely been violated and that this violation serves some sort of purpose, we should

103

identify the work of an agent; an intentional mind acting teleologically.

Indeed, Hume realizes that agency is a fundamental requirement in the definition of a "miracle" since a person possessing an uncanny power to alter natural events would itself be a violation of the laws of nature, even if those alterations were not in themselves contrary to our experiences. This leads Hume onto an extended definition of "miracle":

> "A miracle may be accurately defined, a transgression of a law of nature by a particular volition of the Deity, or by the interposition of some invisible agent." (p. 112, note 4)

We can now observe that our concept of a miracle has a dual nature, and any event must fulfil both requirements if it is truly miraculous; it must violate the laws of nature resulting from the will of some purposive mind.

Furthermore, it is self-evident that any agency that bypasses the usual processes that govern the world cannot itself be governed by those processes; it cannot be natural. Therefore, such an agent would be supernatural, and to perceive any miracle it effected would be to discover evidence for that supernatural being. Thus any incidence of miraculous events would provide evidence for the existence of deities or other supernatural spirits.

Lastly, assuming this definition, a miracle is only possible in a world with at least *some* mechanistic physical laws. Suppose a world in which everything (including the most mundane occurrences) is caused by the direct intervention of deities. Crops fail because the gods will it, storms hamper naval voyages because of a lack of religious appeasement, and so on. We are certainly familiar with similar states of affairs in the records of ancient history.

What would it take for us, in such an environment, to declare that we had just witnessed a miracle?

We declare that a miracle has occurred when we are convinced of the intervention of a supernatural agent to transgress the laws of nature. However, in this alternate world, nature seems to be governed by gods themselves; there are no "laws" (or very few) of the sort that the scientific revolution has revealed to us. Therefore (in the eyes of the ancients), when Odysseus encounters a storm at sea, it is neither a natural event nor a miracle. If Poseidon creates treacherous conditions he does so not by violating any laws, since they are the agents in control of the weather. After all, in the world we are imagining, extreme weather is caused by the volition of the gods. Therefore, according to Hume's definition of a miracle, such a storm is not a miracle. This would apply to any act of divine agency in an environment governed by the whims of gods.

Proportioning Our Beliefs

Hume's argument is concerned principally with the role of testimony. Since we are rarely direct witnesses to miraculous occurrences, usually the only way we could have knowledge of such an event is from the assurance of those present. Should we believe them? If we know them to be generally reliable, if their story is consistent and detailed and if the stories of co-witnesses are successfully corroborated, we might be inclined to take them at their word. However, a miracle is a special sort of event, and we are never (or at best very rarely) acquainted with the transgression of nature's laws. Furthermore, there are plenty of historical examples where the attestation of a miracle is supported by an apparently flawless testimony, and yet we are still inclined to reject them. Even if we accept any particular miracle, its acceptance would entail the rejection of theologically incompatible miracles also supported by strong testimonies; this idea forms a subsidiary argument of Hume (pp.116-7).

Hume's argument rests on the principle that we should "proportion our beliefs to the evidence" (pp. 108). A testimony is evidence for its content, its quality limited by the integrity and discernment of the witnesses. As Hume rightly points out, our past experience affords us counter-evidence to any miracle. Since a miracle is (by a necessary part of its definition) a contradiction of the laws of nature, we must suppose that we have an unbroken series of contrary experiences to the alleged miracle. This is, as Hume puts it, "a full *proof* ... against the existence of any miracle" (p.111). We are left with evidence both for and against the truth of the testimony, and must believe whichever is the stronger, in proportion to the difference in strength. Therefore, we should only believe in the miracle if we have a superior proof *for* its existence.

A testimony alleging a miracle, therefore, would have to be so reliable that if it turned out to be false we would declare it a greater miracle than the one it proposed. This, in a nutshell, is Hume's argument.

We can make the same point in a different way. If someone who usually provides reliable testimonies (say 90% of the time) asserts that a miracle has taken place, then it might be said that the probability of that claim being true is also 90%. This is not so. We must also consider the prior probability of the claim itself; in the case of miracles this would be very low indeed (there are no miracles in ancient mythological worlds such as the one outlined above, so in a law-abiding world such as ours miracles are by definition extremely rare). The 90% hit-rate of this testifier, therefore, would yield far more false-positives than true-positives, and therefore even apparently reliable testimonies are much more likely to be false.

The Problem with Agency

The force of Hume's argument is most obvious when we consider what it would take for us to accept a testimony claiming that a miracle had happened. I will discuss the examples Hume gives: a worldwide darkness, attested by all cultures and traditions, and the resurrection of Queen Elizabeth; the former he might accept while the in latter he would deny that she was actually dead before returning to the living. I agree with Hume here, we should believe the first example and doubt the second. Why this is so is not immediately apparent. Knowledge of both events arrives by the testimony of witnesses, and we can rightly say that we have no experience with either a worldwide darkness or the resurrection of a human being.

We mentioned earlier that agency was a necessary component of a miracle. In the worldwide darkness example, it is unclear what role agency has to play. *Cui bono?* Moreover, the testimony is very strong, as Hume puts it: "extensive and uniform" and "rendered probable by so many analogies" (p.123). While we do not have an experience of the exact nature of this claim, we have similar examples of natural events that might be analogous. For instance, the observation of a bright supernova in 1006 AD, the sightings of comets, or the short-term darkness caused by solar eclipses. Our scientific endeavors have illuminated the causes of these phenomena, and we can be sure that the testimonies, although exhibiting ignorance and superstition, were describing a real event. Therefore, it seems more probable that the event really occurred than that everybody on the planet conspired to fabricate it, or that they were somehow mistaken (this would perhaps be *miraculous*). However, it is not clear that this is really a *miracle*, since it does not appear to be the volition of an agent

The resurrection of a mortal person, as in the second example, is clearly the work of an agent. Not only does it show that Queen Elizabeth is favored, but after the event she will likely command additional respect and reverence amongst her subjects. No matter how strong the testimony appears, Hume would remain unconvinced: "the knavery and folly of men are such common phenomena, that I should rather believe the most extraordinary events to arise from their concurrence, than admit of so signal a violation of the laws of nature" (p. 123). Proportioning our beliefs to the evidence, therefore, we rightly judge that the falsification of the testimony is puzzling but non-miraculous, and we should disbelieve the miracle as a result. Furthermore, now that we have an example where agency is clearly involved, that agency could just as well be the "knavery and folly of men" rather than the act of a supernatural deity. It is clear just who benefits, and why there might be a conspiracy. This is true of any example which, as Hume puts it "[forms] the foundation of a system of religion" (p. 122).

In the first example, although we might believe the testimony, we were not hearing about a miracle as there was no agency involved. In the second example, we understand the testimony to be regarding a miraculous event; however, it seems more likely that the agent involved is very human, and therefore the testimony should be dismissed. In short, the fact that a miracle requires an agent severely weakens any testimony asserting its existence. It seems, therefore that one who would believe in miracles has a dilemma. Either they must deny that agency was involved, and what they describe is no longer a miracle, or they must claim that it was effected by an invisible and supernatural agent, giving us extra grounds for suspecting their testimony.

Hume's argument is a powerful objection to claims that a miracle occurred, but we might also go a step further and say that it also provides an objection to testimonies about paranormal events. Now, many paranormal events are (hypothetically) testable and repeatable. If somebody claims to have extra-sensory perception, then we can test this ability in a controlled setting, to see if they are really able to do what they claim. Claims about miracles do not allow for this, as they are typically one-time events.

There is, therefore a disanalogy between strictly miraculous events and typical paranormal events. Nevertheless, it seems that *some* reports of paranormal events, such as ghost sightings, or any other claim that asks that we believe it on the basis of testimony, might suffer from Hume's objections. It is unclear what role agency has to play in many paranormal occurrences. For example, are ghosts part of the intention of a deity? However, if we "proportion our beliefs to the evidence," we still end up with overwhelming "experimental" data (i.e. our ongoing experience of the world) that these sorts of events do not occur. Further thought on this topic; the relation between paranormal events and the agency of a purposive mind, might be a productive and interesting project. It may perhaps shed more light on how best to approach the testimonies of those who claim to have witnessed a paranormal event that they cannot reproduce in a controlled environment.

Conclusion

Hume is right to disregard any miracle that would be the "foundation of a system of religion." It seems far less miraculous that a few people should lie (or be mistaken) than that a genuine miracle occurred. We are not acquainted with the transgression of the laws of nature by the volition of a deity, and yet we can find many examples where we have received false testimonies. Hume's argument justifies us in disbelieving the well-attested miraculous events we read in ancient and modern texts, without the necessity to provide counter-evidence.

Bibliography

Hume, D. (orig. pub. 1779) *Dialogues Concerning Natural Religion,* Popkin, R.M. ed. (2nd ed., 1998). Cambridge: Hackett.

On Doubting the Existence of Free Will, and How It Can Make the World a Better Place

Jonathan M.S. Pearce

Free will is something that is taken for granted by the average layperson; it is something vehemently argued for by the average religious believer and denied by the average neuroscientist; and it is something deeply cogitated upon by the philosophically inclined legal policy-maker. Whether or not we possess free will has been the subject of debate, hotly contested, over several millennia. In the context of this book, I will be presenting an account of why free will is incoherent from both a philosophical and scientific standpoint. More importantly, though, I will be appealing to the idea that such skepticism over one of the most fundamental components of everyday life is not only philosophically coherent, but is also beneficial to the skeptic who denies it and, arguably, the society which recognizes its falsity.

I will be defining free will in its classical form, a point which has itself caused much ink to be spilled. Accordingly, I will define free will as *the theoretical ability to do otherwise in any given situation.* I say theoretical because, in reality, we are unable to rewind life to a point of seeing whether we could have done otherwise in that same situation. Someone who believes in this believes in what I will label Libertarian Free Will (LFW) and I will call them a free willer (LFWer).

Let us put this in the context of a real world situation. Imagine a person, Joe, at 9.00 on Tuesday morning in this world (W1). Now Joe decides to pick up the phone to phone his mother. On the free will model and the deterministic model, this can be written as: In W1, at t=1, in causal circumstance C, A does X, where causal circumstance is the set of variables in the world at any given time. The causal circumstance might be any and all learning experienced by the agent, their genetic and biological make-up and the environment at the time of making the decision (every single atom and sub-atomic wave function of the universe being in a given position). What it means to believe in free will, then, is that if we continued life for ten minutes and then rewound back to 9.00am, then Joe *could*, in theory, choose differently to not pick up the phone, and make a cup of tea instead. Or, in W2, at t=1 and at causal circumstance C, A does Y. I use W2 here to be an identical world where everything is the same, including the causal circumstance, but is differentiated so we can compare the choices. In reality, they are the same worlds which theoretically diverge at t=1—the moment of making the choice.

The problem for the free willer is that they must argue that in an identical situation, a person has enough reason to choose X and enough reason to choose Y. Or in more logical terms, in W1 at t=1 and in casual circumstance C, an agent chooses X and not-X (or ~X). Both W1 and W2 are identical, and yet they deliver different outcomes! This appears to break the *Law of Non-Contradiction*.

From a different angle, we can simplify causality, the notion of cause and effect, to a causal chain (though it is, in reality, more complex than this). We could say that A causes B, which causes C, D, E and so on. Moreover, the chain regresses back past A to what must be a first cause of sorts: the beginning of the universe or similar. But for a free willer to argue for LFW, it means that they cannot let the causal chain regress—there cannot be *causal reasons* why something happens because this implies determinism[1]—the antithesis of free will. What does this mean? Well, the agent has to be what is known as the *originator* of the causal chain, of the decision. This means, in the same way that theists claim that there must be a First Cause for the universe[2] (God), the LFWer must claim that *the agent* is the first cause for the causal chain of a freely willed decision. But the mantra invoked in the formulation of such a *cosmological argument* is *ex nihilo nihil fit*. Or, from nothing, nothing is made. And only God can do this in creating the universe out of nothing, apparently. But the LFWer contradicts this by claiming that an agent can create a causal chain out of nothing. There can be no regression backwards, because this, as mentioned, implies determinism! This is often known as the Dilemma of Determinism. Either the first cause is random, or it is caused by some other determining factor(s); but for the agent to be the originator of the causal chain it implies that there is no grounding of causal reasoning for the causal chain. This, essentially, equates to the notion that the beginning of the causal chain (i.e. the decision) is arbitrarily grounded. How can an agent have any kind of moral responsibility for a decision which arbitrarily originates out of nothing? In the context of Jim, deciding to pick up the

---

[1] The position of determinism, for the intents and purposes of this essay, is synonymous with incompatibilism—that free will and determinism are not compatible, but that determinism might not be true as far as aspects like random or quantum indeterminacy are concerned. Whilst these might invalidate strict determinism, they do not allow for free will. Thus I am using determinism as a term to cover positions which deny free will.

[2] This is something which is known as a cosmological argument, the most famous of which is the Kalam Cosmological Argument, and one with which I am mildly obsessed. There are problems with cosmological arguments which mean that it is not so easy to posit that God is the First Cause, despite what theists claim.

phone was as a result of him not having spoken to his mother for some time, knowing she was ill, having respect and care for her due to her excellent parenting, it being Mother's Day, having a solid moral foundation, seeing an advert on TV where someone is phoning their family and so on. The LFWer must believe that these causal elements start with Jim—that there is no regress. We cannot map things back (so that the parenting of his mother was caused by her own loving upbringing and having lost her first two children, a supportive network, good moral education herself and so on, all of which can be themselves mapped back further).

Philosopher Paul Russell sums this dilemma up well:

> One horn of this dilemma is the argument that if an action was caused or necessitated, then it could not have been done freely, and hence the agent is not responsible for it. The other horn is the argument that if the action was not caused, then it is inexplicable and random, and thus it cannot be attributed to the agent, and hence, again, the agent cannot be responsible for it. In other words, if our actions are caused, then we cannot be responsible for them; if they are not caused, we cannot be responsible for them. Whether we affirm or deny necessity and determinism, it is impossible to make any coherent sense of moral freedom and responsibility. (Russell 1995, p.14)

One of the most common defenses of Libertarian Free Will (or contra-causal free will) is what I sometimes term the 80-20% approach. Most people, to some degree or another, accept that our lives are at least somewhat, and in most cases, a good deal, influenced. This may be by genetic, biological or environmental factors. But it is hard to deny that, at the point of making a decision, we aren't having our decision influenced by external or internal motivators. This is expressed often as a claim like "Well, we are influenced quite

114

a bit, but we still have some degree of free will" or "I think we are 80% determined, but 20% of our decision-making is freely willed."

Well, when people claim we are, say, 80% determined, but that 20% of an action is still freely willed, we have *exactly* the same problem as stated earlier in this piece—we have just moved that argument into a smaller paradigm, into the 20%. Assuming that we forget the 80% fraction (which is determined) as not being of interest to the LFWer, we are left with the 20%. But this is devoid of determining reasons. So what, then, is the basis of that 20% in making the decision? The agent cannot say, "Well my genetically determined impulses urged me to A, my previous experience of this urged me toward A, but I was left with a 20% fraction which overcame these factors and made me do B" because he still needs to establish the decision as being reasonable. So if that 20% is not just random or unknown (but still grounded in something) and had any meaning, then it would be reasoned, and thus determined! The two horns of the Dilemma of Determinism raise their ugly heads again. We are left with causally reasoned actions or actions without sufficient causal reason, neither of which give the LFWer the moral responsibility that they are looking for.

Having provided this as a rather concise overview of some of the philosophical considerations to make whilst cogitating over whether one has libertarian free will, it is worth noting that there is a plethora of academic literature debating as to whether a) we do or do not have alternative possibilities; b) whether denying alternative possibilities also denies moral responsibility or not[3]; c) whether other definitions of free will suffice to account for free will adequately and whether they, too, allow for moral responsibility. The scope of this essay does not lend itself to such a critical analysis. Suffice to say that I find that philosophical analysis leads me to a denial of free will as defined here. As Derk Pereboom, a philosopher interested in this discipline, states in his book *Living Without Free Will*:

> Obviously undermined is the natural view of ourselves as agent causes, according to which we are not restricted to choices that are alien-deterministic, partially random, or truly random events. This means, in my view, that a pervasive conception of ourselves as agents is lost. Our actions do not result from the sort of power of agency that many naturally believe to be their source, a power that is indeterministic and features a robust capacity for control. (Pereboom 2003, pp. 135-136)

With a philosophically grounded denial of libertarian free will being a given, even if only for the sake of argument here, let us look exceptionally briefly at the scientific evidence to support such a thesis (at least to the point of negating free will[4]).

I will give a brief overview by selecting what can only be described as the tip of the tip of the tip of the iceberg of

---

[3] Harry Frankfurt devised thought experiments which looked to provoke questions about PAP and whether moral responsibility could be attached to actions within this context. A huge amount has been written on Frankfurtian derived ethical philosophy.

[4] For example, randomness or lack of deterministic qualities at sub-atomic level (quantum indeterminacy) may hold, but this does not allow for free will, since the agent still does not have control over these variables.

scientific evidence which points toward a determined causal order.

First, let us briefly consider the famous Benjamin Libet experiments, but with many caveats covering them. These have been replicated and improved upon, but the basic form of them is that test subjects were hooked up to ECGs and ask to press buttons. What he and his colleagues found, to cut a long story short, is that before the subject "decided" to press the button, the non-conscious brain had already kicked into action (this was labelled the Readiness Potential). This seemed to show that the non-conscious brain had "decided" to carry out an action before the conscious brain "decided" (or that the conscious brain was the epiphenomenal by-product of the non-conscious brain). Epiphenomenalism is an interesting theory that consciousness is a natural by-product of the machinations of the non-conscious brain much like steam is a natural by-product of boiling a kettle. However, it is merely a reflection of the non-conscious machinations, rather than being the driving force behind them.

This is an oft-cited piece of research and is also often criticized (by theists looking to defend free will), and there is some credence to be given to such criticisms. First, these experiments seem to refer to our more automatic decisions that we have rather than long-term planning and goal-setting (such as deciding when I go on holiday and where, or what my five-year plan might be). These long-term goals might be freely decided and, in turn, inform our more immediate actions. Also, there have recently been studies that call into question the brain activity. For example, Schurger et al. (2012) seem to imply that the Readiness Potential is derived from spontaneous neuronal activity[5], but this spontaneity does not particularly allow for a free will thesis to flourish. Either way, there is much prevalent research which shows that the sheer volume of work done by the non-conscious brain is staggering in comparison to the conscious brain. Who is really in control?

Genetics and genetically derived behavior are less prone to critical attacks. In the 1960s, Walter Mischel (whose experiments have often been revisited) devised an experiment to look at delayed gratification. His subjects were four- to six-year-old children who were sat opposite a tester who had a plate with either a marshmallow or a cookie. The children were told that the tester would return after some time and if the marshmallow (or cookie) was still there when they got back, then the subjects would get two. Thus, delaying their gratification would ensure that the subjects were rewarded. Already, by this age, children varied widely in how they coped with this scenario. Some used delaying tactics such as distracting themselves, and others simply went in for the kill straight away, and others everywhere in between. The interesting part of the experiment was somewhat accidental. It was supposed to test the age when delayed gratification developed, but with some later follow-up studies in the 80s and 90s, the experimenters found that those would could delay gratification were those who scored better SATs, and who were described by their parents as being "significantly more competent." The skill appears to stay with the children for the duration of their lives and has its basis, it appears, in the biology of the brain (prefrontal cortex and ventral striatum, for example).

---

[5] "[T]he precise moment at which the decision threshold is crossed leading to movement is largely determined by spontaneous subthreshold fluctuations in neuronal activity" is hardly allowing for free will.

In the same way that Mischel found out that adult achievement could be predicted at age four, Gao et al. (2009) have found out that poor fear conditioning at age three accurately predicts criminality at age twenty-three. What this means is that children who were unresponsive to fear (underscored by amygdala and ventral prefrontal cortex dysfunction) at the age of three were far more likely to go on to commit crime in their adult life. Both of these studies show that we appear to be set along a course from birth which our conscious will seems unable to steer away from.

We appear to have our phenotype (traits and characteristics) determined (at least hugely) by our genotype (genetic blueprint). This is why we always seem to be fighting not to be like our parents, and more often than not, losing. As they say, the apple does not fall far from the tree. One can often look at difficult children and either see how they take after their parents or understand their problematic behavior from seeing their present living and social environment. This is often remarked upon anecdotally, and it has been my own observation throughout a career in teaching.

Furthermore, many children who struggle with aspects of autism can have difficulty empathizing. Empathy, I would argue, is the key to morality. When a child who picks up a pencil and throws it into the eye of a child sitting near them is asked why they did it, and they shrug and are unable to give an answer, it is partly because they don't know themselves (they often have a lack of vetoing procedure in their brains) and more often than not because they have an inability to put themselves in the other person's shoes. This skill of empathy would enable them to realize what it would feel like to be the other person receiving a pencil to the eye, and conclude that it would not be pleasant, that they would not like it done to them, and thus they should not do it themselves. Empathy underpins the Golden Rule (do unto others as you would have done unto yourself). But with some autistic children (one cannot generalize as the spectrum is huge and varied), psychopaths and potentially simply "unkind" people, it is more than likely that they have dysfunctional brains which cause a lack of empathy.

Mirror neurons have been posited as the mechanisms in the brain which produce feelings of empathy. These neurons fire when you see someone else doing something, and they produce a feeling in *your* brain that *you* are doing it. For example, if I drive past a car crash at the side of the road and there is blood, and a man in utter anguish and pain crying out, I *feel* that. My mirror neurons fire, and I stop, put him in my car, irrespective of blood and cost of car cleaning, and drive to the hospital with him, even if my car might cost £100 to clean. However, I might get some junk mail through the door the next year featuring a charity and starving children in Africa. My £50 could literally save two children's lives, and yet because the stimuli (small pictures on a piece of junk mail) aren't so conducive to firing those neurons, the empathy isn't induced, and action not taken. And yet on paper, deciding over each action is less obvious. It is theorized that there is a connection between dysfunctional mirror neurons and autism and a lack of empathic behavior.

Similarly, there is a connection between having the "warrior gene" and psychopathy; that many psychopaths have a variant of this gene which promotes certain levels of enzyme release which then appear to affect mental health issues, including psychopathy. It is well worth researching if you are not well-versed in its discovery.

There are other factors which influence kindness. Sometimes called prosocial characteristics, kindness can be genetically influenced. Reuter et al. (2010) have discovered that a gene variant of the COMT gene influences altruism in the form of willingness to donate. This activity of kindness was twice as evident in people with a certain variant as with others. As I evidenced in my own book, *Free Will? An investigation into whether we have free will or whether I was always going to write this book:*

...according to Masahiko Haruno of Tamagawa University in Tokyo, along with Christopher Frith of University College London.[6] They have found that generosity (also seen as the desire for fairness) appears to be automatic—activated by the area of the brain that controls intuition and emotion. The findings are in line with current neuropsychological understanding, that humans are on a scale of being 'prosocial' or 'individualist', with prosocial people being more inclined to share with others. The new research has found that experimenting on two groups of people (a pre-sorted group of prosocial people and one of individualists, sorted by standardized behavioral tests) has produced results that show activity in the amygdala region of the brain increased significantly in prosocial people when dealing with unfair distributions of money. The amygdala did not show such activity in individualists. The more that the prosocial people disliked the distribution of money, the more the activity fired in the amygdala. There are two crucial parts to these findings. Firstly, it was originally theorized that generous people used their prefrontal cortex to suppress selfish feelings, and thus people were thought to control their kindness in the active suppression of selfishness. This experiment showed no activity of the prefrontal cortex, and no difference between the groups in this area. Secondly, the region that was activated (the amygdala) is an area that responds automatically, without thought or awareness.

The researchers consolidated their findings by creating extensions to the experiments, whereby they gave the participants memory tasks to carry out whilst rating the

---

[6] Haruno et al. (2009)

splits. The parts of the brain usually responsible for deliberation over things were being used, and yet the prosocial participants still had the same brain activity, showing that they were not suppressing selfish desires, but responding automatically. Whether the automatic response is in-built or as a result of learned behavior is not important to the determinist, since at the point of making the moral decision of kindness, those automatic reactions are out of one's control, regardless of how and why.

Carolyn Declerck, a neuroeconomist at the University of Antwerp, Belgium, claims that this evidence backs up her own as yet unpublished research that shows that prosocial people are driven by an automatic sense of morality: "So far, all our behavioral and fMRI experiments confirm that prosocials are intrinsically motivated to cooperate." (Pearson 2009)

As the New Scientist declares in relation to Haruno's work:

Haruno will next try to figure out how this difference in the activity of the amygdala arises. It's partly genetic, but also likely influenced by a person's environment, he says, particularly the social interactions during childhood. He says it is interesting to think there might be ways to promote this activity to "realize a more prosocial society." (Pearson 2009)

The part of the brain responsible for moral deliberations is known as the right temporo-parietal junction and moral decisions (qua morality) can be adjusted by giving the brain tiny magnetic pulses known as transcranial magnetic stimulation (see Young 2010). There are clearly at least large aspects of morality that are rooted in the biology and chemistry of the brain.

In all reality, the list of genetic and biological influences on our personalities and actions is nigh on endless. For the necessity of conciseness, I will refrain from listing any further citations and works; suffice to say that the position of determinism, or at least of lack of free will, is rather easy to support with empirical evidence. The idea of libertarian free will, on the other hand, has no supporting evidence apart from "Well, I *feel* like I have free will."

Sometimes I must admit that I get a wave of nausea when I think of humanity's lack of free will, when I imagine the world as one giant irreversible machine. This is a very similar feeling to when I consider why there is something rather than nothing. And by nothing, I mean the utter nothingness of a philosophical nothing. I think that this feeling is natural. However, when I have calmed down to a more pragmatic state, I start considering aspects of free will within a context of where the world is at, and where I would like it to go, which itself has obvious moral dimensions.

The way I think has certainly changed in light of deterministic ideas. Even down to the most mundane of situations. Nowadays, I have to find the reasons why things happen—I am obsessed with causality. For example, I was in a soft play center the other day with my twin toddlers and there was a ball chute which sucked plastic balls up through a mazy tube and into a basket high up in the air. Every so often, the basket tipped up and rained plastic balls onto the children's heads. Because I cannot now dismiss why things happen, I had to stay there filling this basket up with balls until I could decipher whether the basket was tipped by a weight trigger or a timer. I was the geek dad in the corner, obsessed with finding out *why* things happen. Meanwhile, my toddlers were running riot, no doubt.

And this desire to understand causality stretches into a more social framework. Where society is considered, moral responsibility and criminality are never far away from discussion.

So now we have a series of questions:

- Given a lack of free will, do we have moral responsibility?
- Are determined agents morally blameworthy or praiseworthy?
- How do we approach crime and punishment?
- How do we shape our society?

If there is no choice involved in an action, my position (and bearing in mind that this question is and has been a book-length discussion) is that in most senses of the term *moral responsibility*, a determined agent does not have it. However, it seems to me that to apportion moral responsibility is a rather futile pastime. Daniel Dennett opines:

> Of course, people can want just about anything, and a yearning for responsibility might arise when one was in the mood for satisfying a purely metaphysical hankering...
>
> Why then do we want to hold people—ourselves included—responsible? "By holding someone responsible and acting accordingly we may cause him to shed an undesirable trait, and this is useful regardless of whether the trait is of his making" (Gomberg, 1978, p. 208). Once again, the utility of a certain measure of arbitrariness is made visible. Instead of investigating, endlessly, in an attempt to discover whether or not a particular trait is of someone's making—instead of trying to assay exactly to what degree a particular self is self-made—we simply hold people responsible for their conduct (within limits we take care not to examine too closely). And we are rewarded for adopting this strategy by the higher proportion of "responsible" behavior we thereby inculcate. (Dennett 1983, pp. 163-164)

Responsibility, as far as I am concerned, is something which has pragmatic uses such that it can be understood as *attributability* to an agent. If I can attribute an action to an agent in such a way that I can use that attributability to inform future behavior, then responsibility has its uses. But when seen in a moral dimension such that one *deserves* praise or blame, then it is less coherent. This is something that philosopher John Rawls referred to as the "natural lottery." Is it right to praise a great swimmer for having massive feet and great muscle composition that they were born with (genetically)? Or a great mathematician who was also dealt a magnificent set of genetic cards? Such praise (and blame) is, in this way, rather misplaced.

So what use is ascribing moral responsibility? How does it help? Praise and blame are useful mechanisms for informing future behavior both in the agent and in others. The value of praise and blame is therefore seen in the consequences to so doing, thus giving the process a consequentialist moral dimension.

As far as moral worth is concerned, a hard deterministic framework still allows for morality and allows for a person to have moral worth. Just as we might say, "That's a beautiful painting" or "That car is great!" we might also say "Jim is a really good guy." This evaluation makes sense without free will even if we cannot decree that Jim "deserved" that valuation at a fundamental level. As Pereboom continues:

> ...perhaps such appreciation is more like the aesthetic sort than is often thought because it does not involve blameworthiness or praiseworthiness, but it is no less moral for that reason.
>
> ...
>
> The hard incompatibilist position implies that human immoral behavior is much more similar to earthquakes and

epidemics than it would be if we were morally responsible. (Pereboom 2003, pp. 153-154)

Whilst we may not have moral responsibility, we can still hold moral worth.

So where does this leave us? Well, in denying moral responsibility, we are left with a system of praise and blame that are not for any intrinsic value. Agents do not *deserve* to be punished in any kind of retributive sense as this is clearly denied under any system which denies free will. However, as mentioned, the consequentialist element is crucial.

In my own life in teaching and parenting, this has been an interesting revelation insofar as I now consider retributive punishment as emotively driven pointlessness; and reward or blame people so that they themselves act in a particular way in the future and so that others witnessing this do so too. For example, a child who has tried really hard and improved their proficiency at a task as a result of endeavor will be praised so that they are inclined to continue such positive behavior and achievement and so that others in the class see this and are themselves modeled desired behavior and resulting attractive reward. This informs the future behavior for all for a better, more positive future.

Determinism, as a position, leads one to a greater desire to find understanding as to why things happen. In the political world, mantras such as "tough on crime, tough on the causes of crime" have become commonplace. As a determinist, I find that I am forever questioning and searching for causal reasons. Whether it be behavior of children or aspects of society that I think need addressing and improving, being a determinist requires that I shed ideas of emotively-laden blame, and concentrate on a proper understanding of causal circumstances. This then provides me with a better idea of how to deal with a situation and how to go about trying to improve a future causal circumstance

126

so that a negative action doesn't take place in the future, or to ensure that a positive one does. It seems to me to be a more efficient way to live.

Criminal behavior and the justice system present some interesting thoughts with regards to determinism. How do we treat criminals and criminality? Punishment can broadly be seen as having three aspects: retribution, deterrence and rehabilitation (such as through moral education). At the end of the day, the essential goal of punishment should be that the action does not happen again; first, with regard to the perpetrator; and, second, with regard to anyone else in society.

It seems fairly obvious that, considering what I have mentioned before regarding blameworthiness, retribution is incongruent with determinism. The vengeful nature of retribution does not sit well with the determinist who does not necessarily blame the perpetrator for carrying out a crime—they are merely acting out their causal circumstance, after all. There is much more that can be said on this subject, but this must be saved for another occasion.

This, then, leaves deterrence and moral education through rehabilitation. For deterrence, on a consequentialist (utilitarian[7]) basis, the good which is derived by punishing someone is derived by the fact that the punishment causes the perpetrator (and others) not to commit such a crime in the future. Obviously we can't just assume this takes place, and research is necessary to calculate optimal incarceration times to affect positive change without being draconian and resulting in disutility. There is also the utility to the rest of society knowing that these people who have committed crimes are being locked up and thus society is somehow

---

[7] Where morality is derived from the greatest pleasure (least pain) to the greatest number of people.

safer. Such theories of quarantine are not victim to criticism due to the adoption of a determinist worldview and hold up well. Just as quarantine is the right thing to do to protect society from a badly diseased person, prison serves to protect society from a serious criminal.

There are issues with a merely utilitarian approach to punishment, as are described in Pereboom (2003, pp. 166-168). Pereboom, however, lays out Daniel Farrell's appealing justification of special deterrence:

> Each of us has the right of direct self-defense... and... indirect self-defense. Furthermore, if each of us has the right of indirect self-defense, each of us also has the right to carry out the threat against the criminal once its condition has been violated. In addition, if each of us has the right of indirect self-defense, then an impartial agency such as the government has the right to issue a general threat to harm unjust aggressors, and also to carry out the threat once its condition has been violated. Thus, our possession of the right to self-defense justifies punishment as special deterrence. (Pereboom 2003, p.169)

What this all means is that we are (arguably) justified in punishing criminals, and this is not affected by a deterministic worldview.

For me, at the heart of deterministic views of punishment should lie rehabilitation. Correcting a criminal's behavior, by creating a better causal circumstance so that the next time a similar situation arises the agent does not carry out the crime again, is clearly a goal. Being a determinist is about understanding *why* things happen and, in cases where unsavory things happen, the determinist seeks to rectify such situations. That said, it is clear from what has been presented here that laying out a moral code is also paramount. One can conclude toward determinism, but what one does with it is defined by the moral code of the determinist. As with all philosophy, every discipline is interconnected. Thus what type of rehabilitative therapy is used (electro-shock therapy or counselling, for example) will be defined by what moral code one subscribes to.

Pereboom, again, sets out where he sees punishment being justified:

> Just as society has a duty to attempt to cure those who are quarantined for its protection, so it has a duty to attempt to morally educate or cure the criminals it detains for its protection. When this is not possible, and a criminal must be confined indefinitely, his life should not be made unnecessarily unpleasant. (Pereboom 2003, p. 186)

I have, due to demands of space, merely begun to set out that these two notions of punishment (rehabilitation and deterrence) are crucial to a determinist's views on criminality and society. Personally, since "becoming a determinist" I have felt a sense of liberation and enlightenment that has meant that I can now concentrate on *why* things happen rather than *who's to blame*.[8] I feel that there is less warrant for anger in this approach.

If society could take on this approach, without becoming nihilistic or fatalistic, then we would become a much more inclusive and understanding society, I wager. This can only be a good thing.

Such equanimity, such mental calmness, if you will, is a real possibility for a determinist outlook. A complete (or better) understanding of the mechanisms and reasons for all

---

[8] There is a line to a song (*Born of Frustration*) by British band James which references this: "Stop, stop talking about who's to blame when all that counts is how to change." This is a great mantra to live by.

events should lead to a lessening of anger (not to say that anger is impotent; it can be a useful tool). Anger can be very destructive, though, and appears to be fed by moral indignation, itself a product of assigning blameworthiness to agents; an incoherent state of affairs on determinism. As Einstein once stated:

> Schopenhauer's saying, "A man can do what he wants, but not want what he wants," has been a very real inspiration to me since my youth; it has been a continual consolation in the face of life's hardships, my own and others', and an unfailing well-spring of tolerance. This realization mercifully mitigates the easily paralyzing sense of responsibility and prevents us from taking ourselves and other people too seriously; it is conducive to a view of life which, in particular, gives humor its due. (Einstein 1982, pp. 8-9)

In one sense it is an unknown as to how society might cope with realizing that we are lacking in free will, but in another it is quite possible that humanity could flourish. Think of all the anger directed at people who are feared and blamed by others for their sexuality, their worldviews, and their lack of religious belief. A more complete understanding of the mechanisms of the world, of human behavior, in this context, could well be harnessed to progress society to a new state of enlightenment.

The other edifying conclusion that is reached from such deterministic musings is that there is clearly no space for a personal and judgmental God, and all the unwanted baggage that goes with such a redundant concept.

Bibliography

Dennett, D. (1984). *Elbow Room*. Cambridge, MA: MIT Press.

Einstein, Albert. (1982). *Ideas and Opinions,* New York: Crown Publishers.

Farrell, Daniel M. (1985). "The Justification of General Deterrence." *Philosophical Review* 94, 3 (July): 367–394.

Gao, Yu et al. (2009). "Association of Poor Childhood Fear Conditioning and Adult Crime." *American Journal of Psychiatry* 167:56-60. Gomberg, Paul (1978), "Free Will as Ultimate Responsibility," *Philosophical Quarterly,* Vol. 15, No. 3 (Jul., 1978), pp. 205-211.

Masahiko Haruno, Christopher D Frith. (2010). "Activity in the amygdala elicited by unfair divisions predicts social value orientation." *Nature Neuroscience* 13, 160–161.

Mischel, Walter. (1958). "Preference for delayed reinforcement: An experimental study of a cultural observation." *The Journal of Abnormal and Social Psychology,* 56, 57-61.

Mischel, Walter; Ebbe B. Ebbesen, Antonette Raskoff Zeiss. (1972). "Cognitive and attentional mechanisms in delay of gratification." *Journal of Personality and Social Psychology* 21 (2): 204–218.

Mischel, Walter; Shoda, Yuichi; Peake, Philip K. (1990). "Predicting Adolescent Cognitive and Self-Regulatory Competencies from Preschool Delay of Gratification: Identifying Diagnostic Conditions." *Developmental Psychology* 26 (6): 978–986.

Mischel, Walter; Shoda, Yuichi et al. (2006). "Predicting Cognitive Control From Preschool to Late Adolescence and Young Adulthood." *Psychological Science* 17 (6): 478–484.

Mischel, W., Shoda, Y. et al. (2011). "From the Cover: Behavioral and neural correlates of delay of gratification 40 years later." *Proceedings of the National Academy of Sciences* 108 (36): 14998–15003.

Pearce, J.M.S. (2010). *Free Will? An investigation into whether we have free will or whether I was always going to write this book*. Fareham: Ginger Prince Publications.

Pearson, A. (2009). "Generosity is natural for kind-hearted people." *NewScientist,* *http://www.newscientist.com/article/dn18311-generosity-is-natural-for-kindhearted-people.html#.U5x0w_ldWSo* (Retrieved October 1, 2010)

Pereboom, Derk. (2001). *Living Without Free Will*. Cambridge: Cambridge University Press.

Russell, Paul. (1995). *Freedom and Moral Sentiment*, USA: Oxford University Press.

Schurger, A., Sitt, J.D., Dehaene, S. (2012), "An accumulator model for spontaneous neural activity prior to self-initiated movement." *Proceedings of the National Academy of Sciences*, October 16; 109(42): E2904–E2913.

Weill Cornell Medical College (2011), "Marshmallow Test Points to Biological Basis for Delayed Gratification." *ScienceDaily*. September 1, 2011. Archived from the original on October 4, 2011, http://www.sciencedaily.com-/releases/2011/08/110831160220.htm (retrieved October 10, 2011)

Young, Liane et al. (2009). "Disruption of the right temporoparietal junction with transcranial magnetic stimulation reduces the role of beliefs in moral judgments." *Proceedings Of The National Academy Of Sciences* http://www.pnas.org/content/early/2010/03/11/0914826107 (retrieved October 10, 2010)

Pseudoarchaeology: Seven Tips
Rebecca Bradley

In the first years of archaeology, it seemed that almost any new discovery could challenge or overturn the scholarly view of the past. Archaeological deposits were as yet unexcavated, chronologies as yet undevised, millions of scholarly words as yet unwritten. In those palmy days at the birth of archaeology, almost any theory, any speculation, was as good as any another.

A couple of hundred years later, the world has changed. Thousands of tons of dirt have been shifted. Whole museums have been filled. Whole libraries have been written. A vast corpus of knowledge about the ancient world has been painstakingly built up, founded on the evidence, and is constantly expanding and adapting as new data come in, or more rigorous theoretical models replace the old. But there are rules now about how that corpus can grow and change; it is no longer a case of *anything goes*.

We know too much for that now. Many beautiful theories and compelling ideas have been thrown out, for the very good reason that they did not measure up to the masses of incoming evidence. That is, the mainstream of archaeology threw them out—the curious fact is, some of them keep popping up again, generation after generation, in new or even just slightly disguised forms. As well, new speculations keep arising, as part of popular culture: ancient aliens, earth mysteries, psychic and new age approaches, lost supercivilizations. These add up to a field that scholars in the trade call "alternative archaeology"—when they're being polite. A blunter term is "pseudoarchaeology."

There is a lot of it about, alas, in best-selling books, websites, and the visual media; the "alternos" can be amazingly prolific. Some of their ideas are extremely attractive, appealing to a deep-seated love in the human psyche for adventure, dark mystery, arcane ancient wisdom. At the same time, and often jostling the alternos' works in the same shelves in the bookshops, there is a solid genre of popular archaeology that follows the evidence-based narratives of mainstream scholars. How does one tell the difference? What criteria can be used to sort the wheat from the chaff? Here are a few tips to keep in mind, including factors that should cause you to raise mental red flags.

1. Authors' Credentials

Alternos complain that potentially valuable ideas from their camp are ignored because their proponents do not have that union card we call a degree in archaeology. I will freely admit there are some PhDs who are idiots, and some alternos who are both brilliant and well-informed. Nonetheless, that bit of paper does guarantee something very important: that its holder will have spent a fair number of stressful, overworked, underpaid years as a student, intimately exposed to the archaeological database, and to its wider context.

The latter will include the history of the discipline, a good grasp of previous research, a crucial grounding in method and theory, and a fair amount about related disciplines: geology, biology, anthropology, etc. That counts for something. Certainly we should not accept uncritically the pronouncements of PhDs, nor automatically throw out the suggestions of those without them—in fact, self-taught scholars, passionate amateurs, and people whose formal expertise is in fields other than archaeology have a more honorable place in the discipline than the alternos would have the public believe. An author's lack of a broad background in archaeology, however, should at least raise a red flag. A guy with a pair of pliers may do a perfectly adequate job of pulling teeth, but I personally would prefer to go to a dentist. The principle is similar.

Of course, many excellent works of popular archaeology are written not by professionally active archaeologists but by science writers and journalists. In these cases, you can get a good idea of where they stand by looking at the company they keep—that is, at the scholars they choose to interview or consult. Which brings us to the next topic.

2. References

Check the back of the book. Are references provided? If the work contains no references or footnotes, your spidey senses should definitely be tingling. But if it does have references, take time to look at them.

How recent are they? If the bulk of them are older than, say, fifty years, be a little suspicious. This is not to devalue the older syntheses, and of course excavation reports are timeless primary sources, but archaeology actually does move on quite quickly. A competent researcher will make sure he or she has the benefit of the latest evidence. There should be at least a sprinkling of recent material in the bibliography.

Are they reliable sources? If the references are mainly to the author's own works, or to works of known alternos, get your red flag ready—some external validation would be reassuring. Many of the "facts" referenced in those works may have been definitively falsified, debunked, or left in the dust long ago, as archaeology trundled past them.

Even references taken from certain mainstream scholars should raise a red flag. For example, a scholar much referenced by certain alternos is the late Egyptologist Sir E. Wallis Budge (1857-1934), an eminent scholar in his time, but a notoriously careless one. His books on Egypt were popular, but also wildly speculative, inaccurate, and out of touch with the extensive research going on at the time. Egyptologists nowadays warn their students against using Budge because he is unreliable—and yet he is a major source used by, for example, the prominent alterno Graham Hancock. When in doubt about a certain reference, it may pay to google the author.

One more thing. If you find yourself reading a dubious claim that is backed up by an apparently unimpeachable reference, you might check to see whether the cited work really says what the writer *says* it says. Not infrequently, you will find that the meaning has been twisted, taken out of context, or otherwise misrepresented.

3. Idiosyncratic scholarship

This one reminds me of the one joke I ever heard my straitlaced Yorkshire granny tell. A fond mother is watching her soldier son pass by, as part of a military parade. "Ah, look," she says proudly, "they're all out of step but my Albert."

It may, of course, happen that some lone genius working away on the fringes of things is correct, and the consensus of the professionals is incorrect; but it is pretty damn rare. If the book you're reading builds its thesis on, for example, the author's own, very own, patented interpretation of hieroglyphs or cuneiform or Mayan iconography or Melanesian mythology or the Second Law of Thermodynamics, an interpretation not shared by professionals in the field, a red flag should certainly go up.

4. Sweeping Analogies

Analogy is *not* a bad word—both alternative and academic archaeologists use analogies—but there must be good reason to think the similarities on which an analogy is based are meaningful, and not just coincidental, convergent, or even illusionary. If a certain work claims to tie together ancient monuments and cultures on opposite sides of the planet, it's time to break out the red flag.

The misuse of analogy used to be rampant in mainstream archaeology—a century and more ago. Superficial, selective analogies were common currency, especially because a dominant theory at that time held that civilization arose in a single area of the globe, and was then carried around the world by trade, culture-bearers, or mass migrations. This single-source diffusionary model was ditched in the face of evidence that civilization arose independently in a number of zones around the world; however, many of the same long-abandoned ideas, supported by debunked and abandoned analogies, misleadingly presented, are still used to promote alternative narratives.

Pyramids are an obvious example. Alternos use the wide spread of large pyramidal monuments across the world as evidence of ancient contact and cross-influence—or, of linkage through the vanished supercivilization, sometimes identified with Atlantis. The Mesoamerican and Egyptian pyramids are the ones most frequently cited, but some alternos will also bring in stupas and Hindu temples, ziggurats, the temple mound at Tiahaunaco, Monk's Mound at Cahokia, and anything else that is large and vaguely pointed.

But are they valid analogies? Archaeologists patiently point out, over and over again, that the monuments cited have little or nothing else in common. They differ in size, materials, construction, age, form, function, evolutionary history, context, and cultural significance. In fact, the only thing all these monuments reliably have in common is that they are bigger at the bottom than at the top, which is not surprising—that is the default way to build very large stone structures with ancient technology. If there were a global pattern of upside-down or sideways pyramids, that would be a different matter.

Sweeping analogies, especially those that cross large tracts of time and space, should raise an enormous red flag; and yet, such analogies, often linking monuments many thousands of miles and thousands of years apart, are a staple of alternative scholars.

5. Parsimony and Simplicity

The past is messy—the simplest explanation is not always the best one. However, complex explanations should only be resorted to when simpler ones prove inadequate, or when the data demand them. For example, any theory that has to posit such exotic agents as spacemen or Atlanteans has to show first that the unexotic agents—the ancient people themselves—could not possibly have produced the observed phenomena; and then the theorists still have to show good evidence for their particular exotic explanation. Alternos commonly assert that certain engineering feats (the long-suffering pyramids, or Andean stonework, or the Easter Island heads) were not possible to achieve without advanced tools or machinery, but in this they are doing no justice to the skills, sweat, and ingenuity of our forebears.

Similarly, you should treat with extreme caution any claims that rely on abstruse mathematical calculations, aka "junk math," particularly if those calculations involve mathematical theory and methods not known to or used by the ancient people in question. For example, the myriad measurements that can be made on a complex monument like the Great Pyramid of Giza can readily be used to derive pi, or the distance from the Earth to the Sun, or the number of years since Moses led the Children of Israel out of the Land of Goshen, or all manner of advanced mathematical formulae, or the end of the world (in 1914)—all of which have been done in all seriousness. This is despite the fact that Egyptologists have excellent evidence for the effective but fairly crude mathematical methods the Egyptians actually used.

6. Weird Logic

Even apart from the reliability of the facts an author brings forward, you should keep an eye on how reliably he or she builds a case. Something that sounds perfectly plausible may, if you break it down, be suffering from faulty links in the chain of logic. A common pattern in pseudoarchaeological works goes like this: something that has been brought forward as *speculation* at one stage of an argument is treated as a *fact* at the next stage.

To illustrate:

Maybe X happened (speculation).

And since X happened (X is now a fact), then maybe Y happened (speculation).

And since Y happened (Y is now a fact), then maybe Z happened (speculation).

And since Z happened....etc. Lather, rinse, repeat.

Mainstream scholars speculate, too, but are careful to let you know *when* they are speculating, and when they are on firmer ground.

7. 'Tude

Attitude toward the other camp can be a dead giveaway. Mainstream works of popular archaeology tend to be about *archaeology*. Among alternos, however, hostility to the mainstream is a recurrent theme, and one which is unfortunately absorbed by many of their readership. The alternos invest themselves with the appeal of the underdog, and the glamour of the maverick. They present professional academic archaeology as the next thing to a priesthood: elitist, dogmatic, pedantic, held back by investment in the orthodox paradigms, afraid to step out of line.

This is simply not true. The last century of archaeology has been colorful and action-packed—healthy controversy, waves of self-criticism, several episodes of fundamental change, the development of whole new subdisciplines, the rise of middle-range theory, and generation after generation of—believe me—fairly hyperactive archaeologists pushing back the boundaries, and looking for new ways to find out about the past.

There is no one orthodoxy in archaeology—there are a number of schools of thought bashing away at the problems of the past, and sometimes at each other. And the process is not the stodgy, dignified, coldly efficient, and somewhat emotionless process envisioned by the alternos. It's a little more like, say, the Three Stooges doing their tax returns: much head-scratching, eye-poking and yelling of contradictory directions, but in the end, a provisional consensus may emerge. No, archaeology does not suffer from stodginess. New ideas are being incorporated all the time.

The irony is, most pseudoarchaeological theories are *not* new. Some of them are mainstream hypotheses that were falsified and abandoned long ago, for good reason. Some of them are repackagings of old pseudoarchaeological theories

140

that have been repeatedly debunked over a period of decades. For archaeologists hearing them for the nth time, they are just "*déjà vu* all over again."

The New World Order Is Coming for You!
Staks Rosch

I know a secret that you really need to hear. As it turns out, there is a group of the richest and most powerful people on Earth who have been plotting for hundreds of years to take over the world and turn the rest of society into slaves. And guess what? They have very nearly succeeded!!

You don't believe me? It's all true. The Illuminati meet at Bohemian Grove and have been doing so for hundreds of years. They are some of the richest and most powerful men (no women) on the planet and yet for some strange reason they just can't seem to take over the world.

They must be pretty competent because they have succeeded in making billions of dollars in various industries and have somehow been able to hide their plot from the public for hundreds of years. Still, for whatever reason, this group has not yet been able to birth their evil plot to completion—but they are close. They are oh so very close.

Call me a skeptic, but I do have to wonder what they are planning to do once they succeed in actually taking over the world. After all, they are the richest and most powerful men in the world, so they already have money and they already have power. What else do they want? I don't know; and oddly enough, those vigilant watchdogs who informed me about the Illuminati's insidious plot also don't have a clue.

These claims haven't really been thought through. The people who propagate these grand conspiracy theories are certainly sincere and many of them are skeptical of equally ridiculous religious claims, but for some reason they don't really view this grand conspiracy theory about secret shadow governments bent on world domination with the same skepticism. At the end of the day, the evil Illuminati amount to the same as the Devil or the Boogieman. They are just an innocuous evil force responsible for all the evil in the world. This force must be fought by the chosen few. Only those who are special and are able to see behind the façade of ordinary life have what it takes to defeat this unseen enemy.

That is really the appeal to the belief in the existence of the New World Order conspiracy, cults, and evangelical religions. We live in a world that for many seems quite ordinary. We no longer worry seriously about our day-to-day survival. Each day we go to work and do a job that is relatively safe, and then we go home to our families until the next day, in which we do basically the same thing. Some people are so desperate to escape the confines of our safe and predictable world that they have invented a world in which they are action heroes of sorts. This fiction has given their lives meaning and purpose beyond the ordinary. They have become humanity's only hope against the evil Illuminati and their plot to bring about a New World Order, in which everyone has become a slave to the powerful elite.

When something bad happens, we often feel that we need to place blame somewhere and yet sometimes there is nowhere that blame can actually be placed. So some people invent a scapegoat. Who could have caused me to lose that promotion at work or get fired from work all together? Who is responsible for the stock market crash and my 401K being depleted? Who can we blame for that new strain of the flu virus? What about AIDS and cancer? Could it be Satan?

For some people—like Pope Francis—that makes perfect sense, but for far too few, the idea that there is some supernatural demon lord/fallen angel is pure silliness. Fortunately, there are some people who reject the idea of Satan. Unfortunately, some of those people still have a need to assign some sort of blame to various evils in the world and in their lives. They need a more tangible boogieman than the

Devil can offer, and that boogieman has become the Illuminati.

Depending on your level of crazy, there are some pretty big differences as to who or what the Illuminati are. Some believers claim that the Illuminati are as described above, a small group—6000 people tops—of the richest and most powerful men in the world: "old money" families like the Rothschilds and the Rockefellers. For others, the Illuminati are faceless stereotypes like bankers, Jews, and Freemasons. Then there are the crazy few who will add shape-shifting reptilian aliens to the mix. Apparently they are the real power behind the future dictators of the world. As ridiculous as that seems, it really isn't any more ridiculous than the belief in Satan and his demons.

What makes this fiction more interesting than your average religion or cult is that it has some kernel of truth to the delusion. There are super-rich people and corporations in the world who do use their money, power, and influence to exploit workers and buy politicians in order to increase their profit margins. Bankers do hoard money and swindle the middle class into loans and mortgages they can't afford. The government does listen to our phone calls and read our e-mails. Cell phones can be used to track our locations. All this is true, but still the grand conspiracy theory fails to stand up to skeptical inquiry.

While many wealthy individuals, politicians, and corporate leaders do meet at places like Bohemian Grove, it is extremely unlikely they are sitting around plotting world domination. It is much more likely that they are networking for their personal interests—sort of like politicians and less wealthy corporate leaders do on the golf course.

Is some of the networking that goes on nefarious and aimed at increasing the profits of the already super rich? Of course; it would be naïve to think that these individuals had purely the best interests of humanity in mind. But at the same time it would be beyond foolish to believe that there is any type of long-term, Moriarty-type plan designed to enslave humanity.

This paranoia about the Illuminati and the New World Order is not just delusional, but it is also dangerous to humanity's future. In fact, it is almost as dangerous as theistic religion in the sense that it is a delusion which promotes a world divided instead of a humanity united. As technology has advanced, the world has become more interconnected. The European nations have formed a union and there is greater cooperation among many other nations. As time goes on, history seems to be bending toward a society under a single government. This is the evolution of humanity as we learn more and more about good government and how best to allocate resources in a global economy.

This is a good thing. We will have fewer conflicts over resources and fewer wars. We will have a greater ability to work together for the common interests of science, medicine, and space exploration and less of a need to spend exorbitant amounts of money preparing to fight and kill each other. This will also foster our global identity as human beings rather than members of national tribes.

The League of Nations was the start of this process and today we have the United Nations. The United Nations is far from perfect but it could serve as a stepping stone for the next step in the evolutionary chain of government. The problem is that these paranoid New World Order watchdogs are so afraid of any kind of world government that they attempt to sabotage any kind of effort toward international cooperation. They continually "warn" that the evil Illuminati are behind the United Nations in an attempt to create this New World Order in which they will rule over humanity as tyrants.

Of course the Bible comes into it too, despite the fact that at least some of these grand conspiracy theorists are

146

atheists. Apparently there is something in the Book of Revelation about all the people in the world giving authority to the anti-Christ or something (Revelation 13:7). This has widely been interpreted by those afraid of a New World Order as "evidence" for their fears. Plus, the story of the Tower of Babel in Genesis makes it clear that the God of the Bible isn't exactly a fan of human beings working together.

This type of delusional thinking impedes the progress of humanity. These people are going around telling other people that world peace is not a good idea because it plays right into the hands of the imaginary Illuminati's evil plot for a New World Order. The "Beast" himself wants there to be world peace, so it can't be a good thing. They might also point out that someone else once wanted world peace... Hitler! Then again, I don't think Hitler had a global democracy in mind when he marched his armies across Europe and Asia.

Aside from trying to prevent global cooperation and world peace, these New World Order alarmists are ready to prevent progress in other ways too. When there is a new medical breakthrough like the H1N1 vaccine, these people shout, "Conspiracy! Vaccines are just another way that the Illuminati will control people." It isn't just vaccines either; they shout conspiracy whenever the topic moves to global climate change, genetically modified foods, and water fluoridation. When new technologies like smartphones are developed, we are warned that the Illuminati will use it to track your every move. Even our currency is suspect as it contains Masonic symbols and secret tracking devices... apparently.

Really, the richest and most powerful men in the world want to know how I spend my day? I find that impossible to believe. The truth is that cell phones can track our locations and that the government can monitor our location by getting that information from the cell phone companies. This is of course an invasion of our privacy and should only be legal with a warrant which specifies who the government is looking for and why. While it is clear the NSA has overstepped their authority on this matter, it is unlikely that this is a plot by the super-rich and powerful to follow my every move. It's just politicians trying to get dirt on political enemies. It's wrong, but the conspiracy behind it doesn't go all that far.

Of course trying to convince a grand conspiracy theorist that there is no plot to create an evil New World Order is only slightly less difficult than trying to convince religious believers about the irrationality of their ridiculous beliefs. I am often accused as being either too brainwashed by the Illuminati to see the "truth," or someone who is part of the secret conspiracy.

Considering that I make very little money and am almost certainly not anyone of global influence, we can rule out the second accusation. As for the first accusation, it might have some merit. I very well could be brainwashed by the Illuminati's subliminal messages hidden in television shows like *The Simpsons*. Oh, did I forget to mention that? Apparently, the Illuminati send out hidden clues to their diabolical plots to being about the New World Order in *The Simpsons*. You don't believe me? There's a YouTube video for that.[1]

Why would they do this? Perhaps they haven't seen any James Bond movies where the evil villain informs the captured 007 of the plan only to have Bond escape the elaborate death trap and use that information to save the world. Fortunately for us, there are a handful of special people on YouTube who have discovered the Illuminati's hidden messages. The Illuminati are so close to taking over

---

[1] See http://www.youtube.com/watch?v=sILriBhJr2M (retrieved November 10, 2013)

the world, after all these hundreds of years, and they would have gotten away with it too if it weren't for those pesky kids on YouTube.

Why Beliefs Matter
David Osorio

As a skeptic, I usually stumble upon fierce faces because I do not happen to believe what someone else does, or, better said, not only do I not believe it, but I also give my best shot at debunking such magical thinking.

This attitude won't make me too many friends in the immediate future, because, well, people hold their belief systems as fundamental part of their lives. And I think that is one of the key issues about being skeptical of all kinds of beliefs that don't have the right evidence to support them.

If you hold a belief as sacred, wouldn't you want it to be *true* as well? Well, usually, people don't make such distinctions and this is clearly problematic.

Beliefs matter, because people act upon their beliefs.

Quick example: Even though there's no evidence of an afterlife, and there's a good amount of evidence that blood transfusions save lives, some parents are more than willing to let their child die before they allow the transfusion of someone else's blood.

They do this, not because they don't love their offspring (they do love them very much), but because they have *irrational* beliefs.

Beliefs Matter To You

"Ohh, but they're just religious nutcases. I don't buy any of that stuff, I'm fine!"—that could be your answer, and in fact, that's an answer I get most of the time from fellow atheists who happen to believe in all sorts of pseudoscience and

alternative New-Age-y lifestyles. Boutique atheists, a friend of mine calls them.

Anyhow, it doesn't matter if you have religious or non-religious beliefs. What matters is whether they're rational, evidence-based beliefs or not. You see, any kind of belief, religious, political, economic, or another kind, has an impact on the way you make choices.

For example, there are some pseudo-documentaries out there that parade themselves about the place, built on the premise of "money being debt" (if that sounds ridiculous, it's because it is as ridiculous as it can get—that's the monetary slogan that economic libertarians use to sell their deceptive ideology; but I digress[1]). If you have ever bought something or collected money that someone borrowed from you, then you know money is not debt, but the opposite, so you'll refrain to vote for the Tea Party.

It can be, and it is, in your best interest to root whatever you choose to believe in facts and evidence.

Let's take an uncontroversial example.

You are in a monogamous and faithful relationship. One day, you get out of work early and you choose to surprise your significant other, so you make him/her a surprise visit. As it happens, you find him/her in bed with someone else.

You may want to *believe* (s)he did not cheat on you. But that belief would only hurt you, because you have evidence otherwise, and if you don't make yourself deserving of respect, people won't respect you; and your partner could get away with it, making it a regular practice to cheat on you and getting into an abusive relationship. And the chances are you're not into suffering and being abused by someone you're

---

[1] For more information on this concept, see this excellent debunking of the concept - http://soundsfamiliar.blogspot.com/2009/02/money-as-debt.html (retrieved 10/03/2013)

supposed to care for and share your life with, right? (In case you're into that kind of stuff, good for you, I won't judge; I just hope that was no obstacle to make my point clear!)

Or you could be of the opinion that you can evade taxes, and that would get you in jail—at least in the United States. Or you could *believe* that you are holding a toy gun, and kill someone close to you. Or you could believe that Freemasons are holding a conspiracy in alliance with Illuminatis, which could make you go nuts in your everyday life.

Your beliefs impact your everyday life. So that's why you'd want them to be as accurate as possible, so you make the best choices.

Your Beliefs Matter to Me

"OK, so what? That's my problem! Why do you care?"—that could be your response. Not so fast.

If you act upon your beliefs, and you happen to hold irrational, faith-based beliefs then this could well lead you to do irrational and harmful things, not just to yourself, but to everybody else.

For example, back with the Tea Party; if you chose to vote for them, and (say) Sarah Palin was to be elected as US President, it wouldn't just be that loony people were in power, but a crazy person would be in the most powerful job there is, with access to nuclear warheads! This would be a person whose grip on scientific knowledge and method is nothing short of embarrassing, and whose resulting political views (think environment, healthcare and gun rights and Jesus' second coming derived from nuclear war) would have lasting negative impact on millions, if not more. One should be skeptical of belief and the people who adhere to them. A crazy person can be dangerous. A crazy person with power is truly scary.

Or, let's just say, you happen to believe there is no real danger in second-hand smoke. Then you would think that you were justified to not care if I was swallowing your smoke. But that is not an evidence-based belief, thus you would not be entitled to smoke wherever you wanted.

Or you could think homeopathy is actual medicine, and apply this false medicine to your own children, dependent upon you to provide adequately for them and to keep them safe.

So, yes, it is a problem when your beliefs and you acting according to them not only endangers you, but others too. The thing is, we have to share this planet, and when dealing with other people you should fact-check your beliefs.

Beliefs Should Matter to Everyone

But let's say you already have evidence-based beliefs and you don't endanger anyone nearby. Why should we care about other people's beliefs in this situation?

Of course, whether something endangers someone else can be subjective. And consequences can sometimes appear invisible, or off the believer's radar. You could be taking part in causing suffering to other people without knowing about it.

Take meditation, for example. In Colombia, we have the *Instituto Colombiano de Bienestar Familiar* (ICBF—Colombian Institute for Families' Wellbeing). They take care of abandoned children and keep an eye out for their safety and health. It has come to my attention that some of these kids, teenagers especially, are subjected to treatments with meditation involved.

If you knew there's no evidence about meditation being effective, you could think they're misguiding children and mistreating them while thinking they are doing them immense good. And you'd be right![2]

You could feel morally obliged to act upon such knowledge and ask the ICBF to stop such treatments and use evidence-based treatments instead and, if they listened to you, you could be making life so much better for a lot of kids and teenagers. In fact, I sent the ICBF an open letter about them wasting taxpayers' money on meditation and got no response, in case you were wondering.

Another example follows. Just by looking at what science has to say, you could spread the word and let people know homosexuality is a preference —just like liking red over blue— hence it is not curable.

You could save a lot of children time and misery by objectively researching the facts, and helping parents avoid sending gay children to "pray-the-gay-away" therapies and the likes. There is a plethora of research into homosexuality. Not only is the science really interesting, but the need to question one's moral preconceptions also follows. If, for example, we can derive from the research that homosexuality has its causality rooted in a mixture of genetics, biology and environmental influence (and not "just a choice"), then is it right to throw blame on to that person? Is it right to "otherize" or dehumanize them? Where do you get your morality from—is it based on fact? Has it been passed to you unquestioningly?

Let's take racism, for example. Nazis used to think Aryans were superior—even though we are all humans, and

---

[2] Check John Horgan's piece on the *Scientific American* website on this - http://blogs.scientificamerican.com/cross-check/2013/03/08/research-has-not-shown-that-meditation-beats-a-placebo/ (retrieved April 10, 2013)

performance or intelligence is not caused by skin color. That's an unscientific stance, and look where irrational beliefs can get us if people really want them to be true. We should certainly question received wisdom and political ideology. Being rational beings in this broadly enlightened age is a privilege which should not be taken for granted.

Just think how much pain and suffering could be and could have been avoided with sensible approaches to knowledge and truth. Skepticism is healthy. Literally.

So here is an easy way to make this world a better place: question your beliefs and those of the ones surrounding you, and always prefer facts over what you want to be true.

Science Denialism at a Skeptic Conference: A Cautionary Tale
Edward K Clint

On Mars right now, a human-built mechanical spider-scientist is driving around and learning the secrets of our sister world after a months-long journey, remote controlled by people many million miles away. Only a few months ago, particle scientists discovered the Higgs boson, which actually explains why anything in the universe has mass. We are fortunate to be alive in this astonishing time.

This makes it all the more jarring to see how science is routinely attacked by subsets of the same group of humans who can harness its power to accomplish such amazing feats. It is easy to recall recent attacks on science and science education: creationist trespasses on biology classes in Tennessee and Louisiana[1], climate change denial, and attacks on the safety and usefulness of vaccinations.[2]

Among scientists and skeptics, many of us have grown to expect this sort of attack on science from certain conservative and religious corners. But science denialism is not confined to the political right, or to religious zealots. Liberal ideology often factors in irrational arguments against genetically modified crops[3], nuclear power[4], vaccines and immunization, and it is the ideology of most people advancing 9/11 conspiracy theories. Nonetheless, if such

---

[1] http://www.nature.com/news/tennessee-monkey-bill-becomes-law-1.10423
[2] http://children.webmd.com/vaccines/features/california-whooping-cough-epidemic
[3] http://www.scientificamerican.com/article/the-liberals-war-on-science/
[4] http://bigthink.com/risk-reason-and-reality/the-historic-roots-and-impacts-of-our-nuclear-fear

denialism showed up at a *skeptic's* conference, surely there would be outrage. At the very least, one would think that it wouldn't be met with thunderous applause. However, that is exactly what happened in November of 2012 at a conference called "Skepticon."

Skepticon's denialism targeted evolutionary psychology, a thriving, if sometimes controversial, social science with roots going back to Charles Darwin. Bashing evolutionary psychology is as fashionable among some circles as denying evolution and climate change is among the very conservative. For example, the *New York Times* published a stinging op-ed called *Darwin Was Wrong about Dating* by Dan Slater.[5] Slater's polemic sought to refute a body of evolutionary psychology on sex and dating. At Skepticon, the critic was internet pundit, self-described feminist and skeptic, Rebecca Watson. Watson is known for her blog Skepchick.org, as co-host of the popular skeptic podcast "The Skeptics Guide to the Universe," and for speaking at secular and skeptic conferences. Watson has a degree in communications, and regularly speaks on science. The charge of science denialism is a serious one, and I will support the claim with a preponderance of the evidence. It is important to note that science denialism is different from mere criticism or skepticism, both of which are very healthy. I will show that Watson engaged in each of the five tactics commonly used by science denialists.

Being conscientiously skeptical is tough. It is easy to be persuaded, often non-consciously, by one's own biases and interests which can be as mundanely comfortable as an old pair of shoes. Watson's Skepticon talk is a sobering reminder that it is not enough to know about critical thinking and

---

[5] http://www.nytimes.com/2013/01/13/opinion/sunday/darwin-was-wrong-about-dating.html

skepticism. It is no insurance against prejudice to wave the skeptic flag or even to have made it your livelihood. We are all vulnerable, and the single best bulwark against deleterious self-deception we possess is never forgetting about that vulnerability. We must remember it, we must embrace it, and we must accept it as part of our world-view and self-concept. What follows is what may happen when we don't.

A Brief Synopsis

Rebecca Watson's talk was called "How Girls Evolved to Shop and other ways to insult women with 'science.'"[6] Watson reviewed a number of purportedly evolutionary psychology claims, generally in the form of their appearance on newspaper websites. Chiefly by way of ridicule and sarcasm, Watson discussed research on sex differences, such as differing tastes between men and women in shopping, casual sex preferences, and in the favoring of the color pink. Watson's talk began with examples of media distortion – for instance, an incident where a scientific paper was misquoted and otherwise mangled. However, her talk was not merely about the media distortion of science, but about the entire field of evolutionary psychology being bad science or pseudoscience. Her conclusion focused on research demonstrating the demotivational effects of perceptions of stereotypical gender imbalances, which she asserted that evolutionary psychology produces.

The main points Watson seemed to want to drive home were that evolutionary psychology isn't science, and that researchers involved in it work deliberately to reinforce

---

[6] http://www.youtube.com/watch?v=r9SvQ29-gk8

stereotypes and to oppress women. Throughout the presentation she made statements like "evolutionary psychologists are trying to..." and "evolutionary psychology requires..." instead of limiting the target to the media or to a subset of the domain. This led much of her audience to assume she was simply referring to the entirety of the field, or to a large majority of it. Some have objected, including Watson herself, that her talk was about "pop" or bad evolutionary psychology only. I will address this objection in the final section "Rebecca Watson responds."

Points of Agreement

Watson correctly pointed out several important things about evolutionary psychology. For example, the media loves to hype and distort the science to sell newspapers, much as it does with other fields such as genetics. Some examples from her introduction are good ones (though she misquotes some of them). Watson also brought up some truly objectionable research. For example, Watson talked about one of the most infamous names in all of contemporary psychology, Satoshi Kanazawa, saying "he's just the worst." I agree. He seems to be a person who thrives on attention, something of a scientific shock jock. Fortunately, he has been most severely criticized by evolutionary psychologists themselves.

Lastly, Watson noted a Stanford social psychology study which shows that "stereotype threat" can be a powerful force in demotivating people. I tend to agree. I have argued for 50% female representation at secularist and skeptical events for this reason and for others. It is important we encourage everyone with an interest in science to pursue it, regardless of gender. I am not sure what this point has to do with evolutionary psychology, however. I am familiar with no research or researcher who maintains that stereotypes aren't capable of being very harmful to society and to individuals who may be discriminated against. The allegations are made, but the case is not.

A Peculiar Sort of Skepticism

Brian Dunning of the website and podcast *Skeptoid* wrote,

> The true meaning of the word *skepticism* has nothing to do with doubt, disbelief, or negativity. Skepticism is the process of applying reason and critical thinking to determine validity. It's the process of finding a supported conclusion, not the justification of a preconceived conclusion.[7]

This is surely well understood by Watson, who calls herself and website "Skepchick," presumably a truncation of "skeptical chick." Now we may ask, how would a true skeptic investigate evolutionary psychology to reach and support the conclusions that Watson has? Before devising such a talk, the first step should be having a firm grasp on the basics of the subject. Since we are talking about a scientific field, we should expect her speak to at least one expert evolutionary psychologist. She might also read one of any number of reviews or overviews from mainstream researchers. To put Watson's presentation in a slightly different light, just think of how silly it would be to call biologists religiously-motivated

---

[7] http://skeptoid.com/skeptic.php

creationists while pointing to Michael Behe[8] and Francis Collins[9] as examples.

Understanding research methods by reading materials on the subject from authorities and shapers of the field would also be sensible, particularly if commenting directly on research programs, as Watson did. Perhaps most importantly, personal assertions should be limited to the depth of one's own understanding. Any other assertions require citations or quotations of experts because Watson is neither an expert nor a scientist in a relevant field. Citations of experts should clearly support her points.

By all appearances Watson failed to do a single one of these things. She seems to have only the most superficial understanding of evolutionary psychology (or the related subjects of anthropology and sociology) and it is not clear that she's read even one scientific paper. There are many reasons to think this. She quoted no material or information during her 48-minute talk beyond what is mentioned in newspapers and other media or publicly available abstracts. While she derided media distortion in one part of the talk, she implicitly trusted media reports for the bulk of it, even citing a magazine sex survey as evidence against peer-reviewed scientific research.[10] Watson made numerous mistakes in content, misrepresented very basic aspects of researchers' work, got many citations and quotations wrong, and demonstrated ignorance of contrary findings and of basic scientific ideas. Ninety of these are discussed in the Appendix.

When professional skeptic investigators like James Randi, Benjamin Radford, or Joe Nickell want to evaluate a

---

[8] http://en.wikipedia.org/wiki/Michael_Behe
[9] http://usatoday30.usatoday.com/news/religion/2009-10-14-PopeNIH_N.htm
[10] See Appendix number 55

claim, even one they probably personally feel is nonsense, such as ghosts or magnet therapy, they talk to the claimant and they bring in expert testimony where relevant. It should be obvious that one needs to listen to a person's claims to fairly evaluate them. Otherwise, there is a great risk of strawmanning that person's views. If Rebecca Watson spoke with a single evolutionary psychologist about her criticisms or concerns, she failed to mention it. This is discordant with professional standards of skepticism: uncharitable, inept, and in bad faith.

Lastly, we know that Watson has no literacy in evolutionary psychology because she admits this herself. At the end of her talk, an audience member asked Watson if there is any "good evolutionary psychology[11]." Watson then threw up her hands in a shrug while saying,

> Proooooobably? I'm guessing yes, but it's so boring, because you can only make it interesting if you make up everything. Because, really, good evolutionary psychology would be more like, "Well, we don't really know what happened in the Pleistocene, and we have no evidence for this, but maybe this. It's not the sort of thing that makes headlines. So if there is good evolutionary psychology, it's not in the media, and therefore, it might as well not exist as far as the general public is concerned.

Setting aside the strikingly anti-scientific (and incorrect) assumption that only sensationalistic lies about evolutionary psychology can be interesting, as well as the jarring ignorance that a scientific field composed of thousands of researchers working for decades and publishing in numerous

---

[11] The answer is yes. See Appendix number 90

reputable science journals only "probably" has some good work being done, Watson clearly revealed that she is only familiar with evolutionary psychology in the media, having just moments before shown how unreliable the media is.

Watson repeatedly cited outliers, people and publications not involved with evolutionary psychology, and disreputable examples of each (as well as a few reputable sources). The first work she mentioned in her talk is important because it sets the tone and is, presumably, important to her thesis that evolutionary psychology is pseudoscientific and sexist. She cited a *Telegraph* article[12] referring to a study done by one Dr. David Holmes about the psychology of shopping. However, this is an unpublished, non-peer-reviewed study conducted by a non-evolutionary psychologist and paid for by a business to help them increase sales, which she nonetheless misquoted, mischaracterized, and ascribed fictional claims to.[13] She spent several minutes on Satoshi Kanazawa, a man widely considered disreputable within evolutionary psychology (Alvergene et al, 2011). Even quoting him, Watson referred only to an interview with a tabloid newspaper. There was no effort to mention peer-reviewed research, or to include what mainstream evolutionary psychologists think.

Supporting the extraordinary claim that a large scientific domain is both generally sexist and methodologically bereft requires extraordinary evidence. Such a claim should entail very serious and careful examination. The evidence should be obvious in the major, reputable works, and not merely scattered and butchered news fodder.

---

[12]http://www.telegraph.co.uk/news/newstopics/howaboutthat/4803286/Shopping-is-throwback-to-days-of-cavewomen.html
[13] See Appendix numbers 1-14

Watson produced no evidence that can sustain her outlandish claims. Even if Watson were accurately representing every person, paper, and claim in her talk, she'd have succeeded only in proving that a small handful of people making claims about sex differences are academically and/or ethically compromised. Her set of evidence is simply too limited to say anything more. The problem is not merely that her claims are often faulty, but that she seems not to understand (or not to care?) what a skeptical inquest requires. She does not seem to know how, or perhaps is not willing, to support her own claims. This is a peculiar sort of "skepticism" indeed.

Rebecca Watson Uses All 5 Tactics of Science Denialists

Summarizing the work of Mark and Chris Hoofnagle, Pascal Diethelm and Martin McKee wrote a paper on science denialism,[14] providing criteria and defining it as "the employment of rhetorical arguments to give the appearance of legitimate debate where there is none, an approach that has the ultimate goal of rejecting a proposition on which a scientific consensus exists." Denialism employs some or all of the following tactics.

1. **Conspiracy theories**
When the overwhelming body of scientific opinion believes something is true, denialists won't admit scientists have independently studied the evidence to

---

[14] http://eurpub.oxfordjournals.org/content/19/1/2.full.pdf

reach the same conclusion. Instead, they claim scientists are engaged in a complex and secretive conspiracy. The South African government of Thabo Mbeki was heavily influenced by conspiracy theorists claiming that HIV was not the cause of AIDS. When such fringe groups gain the ear of policy makers who cease to base their decisions on science-based evidence, the impact on human lives can be disastrous.

2. **Fake experts**

These are individuals purporting to be experts but whose views are inconsistent with established knowledge. Fake experts have been used extensively by the tobacco industry, which developed a strategy to recruit scientists who would counteract the growing evidence on the harmful effects of second-hand smoke. This tactic is often complemented by denigration of established experts and attempts to discredit their work. Tobacco denialists have frequently attacked Stanton Glantz, professor of medicine at the University of California, for his exposure of tobacco industry tactics, labelling his research "junk science."

3. **Cherry picking**

This involves selectively drawing on isolated papers that challenge the consensus to the neglect of the broader body of research. An example is a paper describing intestinal abnormalities in 12 children with autism, which suggested a possible link with immunization. This has been used extensively by campaigners against immunization, even though 10 of the paper's 13 authors subsequently retracted the suggestion of an association.

4. **Impossible expectations of what research can deliver**

The tobacco company Philip Morris tried to promote a new standard for the conduct of epidemiological studies. These stricter guidelines would have invalidated in one sweep a large body of research on the health effects of cigarettes.

5. **Misrepresentation and logical fallacies**
Logical fallacies include the use of straw men, where the opposing argument is misrepresented, making it easier to refute. For example, the US Environmental Protection Agency (EPA) determined in 1992 that environmental tobacco smoke was carcinogenic. This was attacked as nothing less than a "threat to the very core of democratic values and democratic public policy."

Using this pre-existing yardstick hopefully precludes some margin of bias on my part. Note that the authors do not require that all criteria apply to establish one as a denialist, but I will show that Rebecca Watson made use of each of these tactics.

Watson's denialist tactics

1. **Conspiracy theories**
Watson frequently spoke of a general, diffuse, "evolutionary psychologists do..." When she cited researchers by name, they were held as examples of the domain, and not distinguished as an exception. She also often spoke frankly to their devious, immoral intentions. Not just that they're mistaken about their claim or that their method is flawed, but that they actively wanted to oppress women and reinforce harmful stereotypes. According to Watson, they work together toward goals such as defending rape as "natural" and therefore good or the idea that women do not

enjoy sex as much as men (see video indices 20:07, 22:43, 23:41, 35:40, 36:08, 38:40). For example, at minute 36:39: "So now evolutionary psychologists ignore all that [that sex roles changed after the industrial revolution]." So it is "evolutionary psychologists" (note that she did not say "some evolutionary psychologists" and not journalists) are the problem. No evidence was presented that could establish these ulterior motives in such a large group, and as I shall explain, they are entirely false. Mark Hoofnagle wrote the following on ScienceBlogs about conspiracy theories; not Watson's, but his words fit equally well here[15]:

> But how could it be possible, for instance, for every [sic] nearly every scientist in a field be working together to promote a falsehood? People who believe this is possible simply have no practical understanding of how science works as a discipline.

2. Fake experts
Fake experts are not featured prominently in Watson's talk. However, at the end, Watson cites several fake experts whose opinions on the science are inconsistent with established, uncontroversial knowledge, and some of whom also appear to be sources she used for the talk. She implores the audience to read Cordelia Fine's *Delusions of Gender,* a book seeking to justify a radical social constructionist view of gender differences. Scholarly reviews have agreed that Fine makes some good points, but have also criticized Fine for cherry-picking studies as examples that are amenable to her conclusion and ignoring the rest:

---

[15] Mark Hoofnagle himself does not agree with my assessment of Watson's talk as science denialism.

Despite the large amount of junk science on the topic that is reported in the popular media and in some academic outlets, there are also consistent findings of sex differences that hold up across studies, across species, and across cultures. Most of these are ignored by Fine. -Diane Halpern, Science (Halpern, 2010)

However, there is more to the prenatal testosterone research than the few Baron-Cohen studies she mentions and more to the study of clinical populations affected by early testosterone than CAH girls and their play preferences. Fine's selective approach leaves the reader with the impression that much of research into the organizing effects of prenatal testosterone on the brain is invalid and unreliable. In reality, the research in this area is extensive, complex and, yes, uncertain, but not, for those reasons, worthless. The extent of this literature is evident in a review of this research that incorporated almost 300 studies (Cohen-Bendahan et al. 2005). Included were investigations of four different clinical populations, four different direct measures of prenatal hormones, and six different indirect measures. -Margery Lucas, Society (Lucas, 2012)

Watson goes on to recommend the blog website of Greg Laden. Laden is a bioanthropologist who is on record uttering unscientific opinions such as the claim that men are testosterone-damaged women[16]:

The problem with men, as a group...is that at various points along the way on their journey from the female template on which all humans are built biologically, they have been

---

[16] http://scienceblogs.com/gregladen/2012/08/02/men-testosterone-damaged-women/

altered in ways that make them dangerous assholes. Even when we try to reduce the male-female difference as a society, men who do not willingly participate in that often end up being fairly nasty, dangerous beasts; they may be rapists, they may be batterers, they may be some other thing. They break our efforts to have an egalitarian peaceful world. In a way, they are broken. They are damaged, if you will. Some of that damage is facilitated by what you may know of as testosterone.

In July of 2012 Laden articulated the same point, saying, "Just like a male is a broken female, a dog is a broken wolf."[17] Laden's unique views on sex and gender are not representative of sound scientific understanding.

Watson recommended blogger Amanda Marcotte to her audience. Marcotte has written that:

Evo psych evolved to meet the need of the media to have a constant influx of stories justifying sexism through "science." Because it's a whole lot easier to get media attention to your work if your conclusions are that women are (fill in misogynist stereotype) and/or men are slaves to certain sexual signals that make it biologically impossible to treat women as they would someone they considered a full human being. ... "Women are born whores," is quite possibly the favorite sacred belief of evolutionary psychology[18].

---

[17] http://youtu.be/v5uvPhZ29fs?t=38m20s
[18] http://web.archive.org/web/20100613000929/http://pandagon.net/index.php/site/comments/diamond_orgasms_are_a_girls_best_friend/

Evolutionary psychologist Robert Kurzban has called Marcotte's criticism "disgusting," and "ignorant."[19]

3. Cherry picking

Watson spent most of her time referring to stories that appeared in the general media and popular science books. She focused on some of the worst examples she could find, such as the interviews (not publications) with the disgraced Satoshi Kanazawa, instead of focusing on mainstream, reputable researchers. She also generally limited her citations to the sub-topic of sex and gender differences. While it is understandable that she may choose a narrow topic to present to a conference, she frequently made claims about the field in general, not merely as it pertains to sex and gender differences. For example, she cited Stephen Jay Gould's "just so stories" criticism, (long dismantled by biologists[20] and others[21]), but then used only sex and gender claims as examples.

In a 2012 interview, after giving the same talk, Watson said "...I just get so tired of seeing women evolved to shop, women evolved to like the color pink, women evolved to be terrible at math and logic ... and men evolved to rape."[22] Bearing in mind that Watson has said that evolutionary psychologists are doing what they do in order to oppress women, justify rape, and maintain stereotypes, we must assume Watson can tell us how evolutionary psychology hypotheses in other areas such as coalitional psychology, social exchange, language, nutrition and diet, altruism, belief

---

[19] http://www.epjournal.net/blog/2011/10/amanda-marcotte%E2%80%99s-ugly-prejudices/

[20] http://www.bostonreview.net/BR25.2/alcock.html

[21] http://www.stephenjaygould.org/reviews/pinker_exchange.html

[22] https://soundcloud.com/techniskeptic/rebecca-watson

formation, and others all oppress women and support gender stereotypes. Watson ignored the majority of the content in the field she demeans as "not science."

4. Impossible expectations of what research can deliver

Some of Watson's criticisms would un-make many sciences were we to take them seriously. For example, she says "[evolutionary psychologists] never tell us what genes" as if this is a grand indictment of evolutionary psychology. There are scientists making in-roads in this area, but tracing the path from genes to structures to behavior is difficult-to-impossible, except in the case of some diseases and disorders. Further, we certainly don't hold any other sciences to that standard, even the ones for which genes and adaptation are critical. Does anyone know precisely which genes make a cheetah fast, and exactly how they accomplish that? What genes color the peacock's feathers or give rise to the fish's gills? Shall we toss out all the evolutionary biology for which we do not have the genetic bases identified? I should hope not. Cognitive science also focuses on models methodologically divorced from physical stuff like genes and even neurons, but no one doubts that genes and neurons make cognitive capabilities possible.[23]

At minute 15:41, Watson derisively explained her view of the method of evolutionary psychology as picking a behavior, assuming it is evolved, and then finding "anything" in the past that might be relevant to it. She seemed to be balking that such a hypothesis is just totally made up. Yes, Ms. Watson, it is. That is how science works. It is not known

---

[23] See Appendix numbers 64 and 81

what the answers are before starting, so a researcher makes as good a guess as they can and then tests it.

Similarly, at several points Watson criticized exploratory research into new hypotheses because the results vary as methods and hypotheses become refined. She criticized these authors for speculating about explanations of their findings, and for disagreeing with each other. Evidently, Watson's scientific standard is that testing a new hypothesis with a new experimental design should immediately work perfectly and deliver incontrovertible conclusions on which all researchers agree.

At 13:39 Watson said that we can't know enough about the distant past to make assessments of what evolutionary pressures might have existed. She referred to variation in climate and environment and that the lives of our ancestors also "varied." In other words, evolutionary psychologists can't make any assumptions about the past. We can't assume women got pregnant and men didn't, or that predators needed to be avoided, or that sustenance needed to be secured through hunting or foraging. Yet these are *real* and valid assumptions that evolutionary psychologists use. If we were to toss out evolutionary psychology for this reason, we must also toss out much of biology, archaeology and paleoanthropology. Much care must be used in deciding what can and can't be assumed about the past, but archaeologists, paleoanthropologists, biologists and evolutionary psychologists know this well. Watson demonstrably understands this, as she referred to Steven Kuhn and Mary Stine's hypothesis about the social conditions of the pre-historic Paleolithic at minute 14:48. It seems that Watson was prepared to believe accounts of the past so long as they supported her position.

5. Misrepresentations and logical fallacies

Ninety of Watson's errors, false and misleading claims, and other examples of ideologically compromised rationality are listed in the Appendix. Many are variants of misrepresentation and logical fallacies. For the sake of brevity, I will describe just one here.

Watson committed the moralistic fallacy by way of accusing others of its inverse twin, the naturalistic fallacy. To presume *a priori* that what is desirable corresponds with what is found in nature is to commit the moralistic fallacy. For example, "A universe without a lawgiver would have no basis for morality—which would be a ghastly state of affairs—and so God must exist."[24] In this case, since it would be preferable for rape not to be natural—because people sometimes think what is natural is good—we must conclude that it is not. The middle term is the accusation of the naturalistic fallacy, the idea that people will think what is natural must necessarily be good. Watson asserted that "they" (that shadowy, diffuse *they* of evolutionary psychology) enthusiastically commit the naturalistic fallacy and, in fact, that research is designed to dupe and mislead by its use.

She spelled it out at 38:30: "men evolved to rape... it was used as a sort of 'well it's natural for men to rape therefore we don't really need to look into ways that we can change our culture to stop men from raping...'" Who is this *they*? Who is doing the using? Watson did not say but she asserted that if "they" believe rape is about sex, and sex is good because sex is natural, then rape must be natural and therefore good. This is an outright absurdity. It is every shade of wrong from the rainbow of ultimate wrongness.

Scientists also study murder; it does not mean they wish to morally justify murder. Hurricanes and hemlock are

---

[24] http://en.wikipedia.org/wiki/Moralistic_fallacy

natural, but bad. The evolutionary psychology of rape informs that rape is a *more* heinous violent crime than other types of assault, not less so. Commenting on Randy Thornhill and Craig Palmer's book on the subject, John Tooby and Leda Cosmides[25] wrote, "Thornhill and Palmer argue that women evolved to deeply value their control over their own sexuality, the terms of their relationships, and the choice of which men are to be fathers of their children. Therefore, they argue, part of the agony that rape victims suffer is because their control over their own sexual choices and relationships was wrested from them." The conclusion here is that the crime is much more emotionally devastating than the mere violence or trespass indicates; also, it implies that men probably can't understand the true anguish of the experience for women.

In their book on rape, Thornhill and Palmer discussed and refuted the fallacy they're accused of—nine times (Thornhill and Palmer, 2001).

Many influential figures within evolutionary psychology are unpersuaded by the notion of rape as an adaptation, such as Don Symons, David Buss and David Schmitt (David Michael Buss and Schmitt, 2011). This makes a conspiracy view of the field as monolithic sexists rather unlikely. Lastly, Thornhill and Palmer themselves have said the topic is worth studying to help reduce the rate of rape, not to justify it.[26]

Watson never mentioned Thornhill, Palmer, Tooby, Cosmides, Symons, Buss or Schmitt. In fact, she didn't mention anyone at all while hurling damning allegations.

---

[25] http://www.cep.ucsb.edu/tnr.html
[26]http://en.wikipedia.org/wiki/Sociobiological_theories_of_rape#Naturalistic_f allacy

Science Denialism: A Losing Strategy

Philosopher of biology Elliott Sober wrote in his book, *Philosophy of Biology,* about the research program of evolutionary biology[27], known as adaptationism:

> Adaptationism is first and foremost a research program. Its core claims will receive support if specific adaptationist hypotheses turn out to be well confirmed. If such explanations fail time after time, eventually scientists will begin to suspect that its core assumptions are defective. Phrenology waxed and waned according to the same dynamic (Section 2.1). Only time and hard work will tell whether adaptationism deserves the same fate.

The proof of the pudding is in the tasting: creationism, for example, hasn't passed that test. It hasn't given us a better understanding of life, and it hasn't spawned new questions for scientists to explore. Perhaps even more damningly, it has no legitimate body of work to support its own hypotheses. Sober also discussed how the once-scientific phrenology similarly failed as new knowledge was acquired. Science denialists have argued that climate science is akin to the failed phrenology[28], that it is wrong and misguided. This didn't stop the scientists though, who kept working and discovering more and more. Today, climate science is bigger and better than ever. It has innovated and synthesized methods. We've gained incredible new insights about how global climate systems function over time, and especially about the life of polar glacier geo/eco-systems. The

---

[27] http://203.158.253.140/media/e-Book/science/Sober%20-%20Philosophy%20of%20Biology.pdf
[28] http://wizbangblog.com/content/2011/06/19/climate-science-the-new-age-phrenology.php

original findings that the earth is warming have been replicated and supported by new evidences, even if some early research was flawed.

Evolutionary psychology has followed a similar trajectory in recent decades. Roundly criticized in the 80s and beyond[29], researchers were not deterred. Although there are always going to be some poorly produced studies, researchers have weeded out failed hypotheses and have refined methodologies. The influence of evolutionary psychology has steadily grown. Some evolutionary psychology theories, once controversial, are now accepted by mainstream psychology. Every college psychology 101 textbook features a bit of evolutionary psychology (of variable quality, granted). New areas of investigation are being explored which may shed important light on critical aspects of the lives of people, including evolutionary medicine. Michael Shermer remarked on the mainstreaming of evolutionary psychology back in 2009.[30] Despite some real challenges, evolutionary psychology is a science success story. All the nay-saying in the world can't change that. Denialism is a losing strategy. Good science always wins in the end.

Lingering Questions for Watson

Watson appeared to contradict herself in several important ways making it unclear what she believes the answer to these key questions are."

---

[29] http://en.wikipedia.org/wiki/Not_in_Our_Genes
[30] http://www.skepticblog.org/2009/06/02/darwinian-psychology-goes-mainstream/

i)   When did the sexual division of labor begin?[31]

Watson said at minute 14:43, "Recent research by anthropologist Steven Kuhn suggests there was no sexual division of labor prior to the Paleolithic" (Steven Kuhn and Mary Stiner's 2006 paper Watson cited, without crediting Mary Stiner, contains the hypothesis that *Homo sapiens* development of a sexual division of labor 40,000+ years ago allowed them to out-compete Neanderthals). Around 36:08 she remarked, "Prior to the 19th century it was expected that men would retain an equal hand in raising children and helping out around the home... then when the industrial revolution came around men started working the factories leaving women at home."

ii)   What is the standard of evidence, exactly?

Watson recounted the story of V.S. Ramachandran's "hoaxing" of evolutionary psychology around minute 16. In 1997, Ramachandran wrote an article titled *Why Do Gentlemen Prefer Blondes* offering an evolutionary account explaining why men have a sexual preference for women with blonde hair (note that this was just a speculative essay. There were no experiments, no methods, and no findings). Watson incredulously explained to the audience "and it got published" pausing for laughter. She went on to call Ramachandran the "honey badger" of science because he "don't give a fuck" implying his move was some sort of coup against evolutionary psychology. She did not mention that

---

[31] See points 30 and 83 in the Appendix for more discussion

the journal in question, *Medical Hypotheses*, had nothing to do with evolutionary psychology. What's more, the journal purposely invites "radical, speculative" pieces and at the time was not peer-reviewed.[32] Don Symons commented that if it is peer-reviewed, "it must be by chipmunks." Watson invited her audience to take seriously what had been published in *Medical Hypotheses*, a journal that got itself in trouble for publishing an article denying that HIV causes AIDS (Enserink, 2010a; "Guide for authors," 2012). It could only matter that Ramachandran's paper was published in the journal if you thought said journal was "legit."

Even more bizarrely, Watson denigrated a study by a professional psychologist, Dr. Holmes, because it had been funded by a shopping center. Watson said "This is actually marketing disguised as science." Some twenty minutes later (around index 27:40) Watson cited a survey to refute published, peer-reviewed studies. That survey was produced entirely by *Self* magazine for publishing in same and apparently involved no social scientists of any kind.[33]

iii) Are WEIRD subject studies acceptable sources?

WEIRD stands for "western, educated, industrialized, rich and democratic" and refers to common research subjects of social science research which can create an impoverished view of the understanding of human behavior if over-generalized. Watson denigrated three studies she did not approve of as using "white, middle class..." subjects (two of the three times incorrectly[34]). However, Watson favorably

---

[32] See points 34 and 35 in the appendix for further discussion

[33] See points 13 and 55 in the Appendix for further discussion

[34] See Appendix points 38, 43, and 70

cited at least six studies to illustrate her own points, which all had WEIRD participants.[35]

iv) Why so flippant?

Watson's talk was peppered with snark and sarcasm. She seems to have spent very little time researching the topic and doesn't appear to treat the topic seriously. I do not merely mean that she does not take evolutionary psychology seriously—but the entire topic, including her own contentions, is more a comedic performance than an informative presentation.

Watson sees evolutionary psychology as fit for ridicule. She flippantly says that mocking it "never gets old." Even so, what about the impact evolutionary psychology might have? That seems less than amusing. For the sake of argument, let us imagine everything Watson believes is correct: Those who conduct research in the field are frequently misogynists who are dedicating many years to the pursuit of justifying harmful stereotypes and oppressing women. They've succeeded in compromising peer review, and the professional journals which publish them are mouthpieces of the patriarchy whose scientific rigor is limited, at best. They've infiltrated the top universities in the world, and they've established growing departments at said locales and have their own conferences[36] and ever-larger presences at others. They've even succeeded in having much of their literature and research perspective accepted by mainstream social science.

If I believed that all of this was true, I would be horrified. The potential harm to society and to behavioral science

---

[35] See Appendix points 35, 37, 53, 56 and video indices 39:00 and 42:40 respectively

[36] http://www.hbes.com/

would be almost incalculable. Thus if I were to give a talk on it or write about it, I would dig deep. I would cite mainstream sources so that no one could dismiss me as cherry-picking. I would locate reviews of dozens or hundreds of studies, instead of citing one or two in tabloid newspapers easily dismissed as outliers, or taking the word of an author trying to sell books. I would read full published papers and foundational literature, not blurbs from the *Telegraph* about unpublished studies so that my understanding would be robust and accurate. I wouldn't make an unserious, sarcastic tone my main presentational style because the stakes would be so high.

Watson wanted us to believe this great dark power is working, inhibiting social justice, hurting real people and the advancement of science, and that it is entertaining to talk about. She said, for example, that it is working to justify rape. To make rape okay. But hey, no big deal, right? Not big enough to research properly or to stop making jokes for two minutes. This flip attitude lacks empathy, and I find it ethically repugnant. If even close to true, her claim isn't amusing. It deserves real skeptical inquiry and serious investigation and she gave it none of this.

Rebecca Watson Responds

After I originally published a version of this writing in December of 2012, defenders of Rebecca Watson immediately responded by disparaging my character and insinuating that I had compromising ulterior motives.[37] After taking much care to speak to points of substance and to support my

---

[37] http://freethoughtblogs.com/pharyngula/2012/12/03/oh-gob-evo-psych-again/

criticism with evidence, I was saddened by the petty and unconstructive mode of reply. My aim was not to attack Watson, but to challenge a few of her unnuanced views about science and skepticism with which I have professional experience. Had Watson chosen to use my feedback to produce a sound, more sophisticated criticism of evolutionary psychology (entirely reasonable to do) I would have been very glad for it. My motive is ultimately irrelevant to the *validity* of my criticisms here. They stand or fall on the evidence alone, as any skeptic ought to know.

As I write this, it is more than a year from the time I published my critique. Watson's one and only response was a single paragraph comment on a blogger's web page defending her against my critique.[38] In it, Watson conceded to three errors which I had pointed out. Namely, that Dr. Kruger was of the University of Michigan, not Chicago; that the color preference study occurred in England, not in China; and finally, that the "Why People Have Sex" study was racially diverse, not merely "white." Watson also wrote:

> I saw Clint's post but as I'm traveling, I have no time to write anything up, so I'm very glad that you've done a great job of it. I'm actually giving this talk again tomorrow and I'm quite thankful to people who have given me notes and corrections.

Evidently, she had no time to write anything up after travelling, either, as she never mentioned the matter again in spite of being asked to respond.[39] The counter-argument which Watson vaguely praised is that I misunderstood her

---

[38] http://freethoughtblogs.com/almostdiamonds/2012/12/03/science-denialism-the-role-of-criticism/#comment-156490
[39] https://twitter.com/minus1cjb/status/321872266994536448

thesis. It was asserted that her talk was restricted to "pop" evolutionary psychology and not about evolutionary psychology in general. This is a curious defense for a variety of reasons which I will explain, but first note the immediate implication. Watson is a professional speaker, giving many talks each year. She holds a degree in communications. We are asked to believe that she was unable to make her thesis clear, having had 48 minutes to do so. Clarity is doubly important if a topic is controversial, as evolutionary psychology often is. I would invite the readers to watch her talk and decide how "obvious" it was for themselves[40], but I must say that before I wrote my critique, I read what other people thought the talk was about. Here are some of those comments:

> You do not tear down an entire field of research because a small number of papers released (and not even in respected journals) are offtaste.[41]
> —discomcomcobulated

> ... Some of the studies she cited (particularly the casual sex and gender differences) one [sic] appear in my psych textbooks ...and are taught in class. So, no, it's not just silly media talking about them.[42] –Kate Donovan

---

[40] http://www.youtube.com/watch?v=r9SvQ29-gk8
[41] http://freethoughtblogs.com/ashleymiller/2012/11/21/how-girls-evolved-to-shop/#comment-30503
[42] http://freethoughtblogs.com/ashleymiller/2012/11/21/how-girls-evolved-to-shop/#comment-30507

… Watson also shone an unflattering light on evolutionary psychology, which is a discipline with a lot of problems.[43] — James Croft

The female-to-male ratio seemed really good at this conference …, a sense that was reinforced by the crowd's enthusiastic response to Rebecca Watson's speech denouncing the pseudo-science of "evolutionary psychology."[44] —Amanda Marcotte [note the scare quotes]

Rebecca Watson's humorous talk on the sexist pseudoscience rampant in evolutionary psychology was immensely entertaining and got boisterous applause.[45] —Adam Lee

All of the above save the first came from supporters or fans of Watson. Here are five more comments that appeared after my critique and in response to my question regarding her actual topic, including ScienceBlogs's Mark Hoofnagle:

[Watson] did attribute these findings to evo psych as a whole and that is somewhat unfair.[46] —Mark Hoofnagle

Although Rebecca's examples were of pop psychology and the media presentation of research (both genuine and motivated "research") she was clearly aimed [sic] her

---

[43] http://www.patheos.com/blogs/templeofthefuture/2012/11/skepticon-5-day-two-rebecca-watson-on-evolutionary-psychology/

[44] http://www.rawstory.com/rs/2012/11/13/skepticon-demonstrates-that-pro-fun-means-anti-harassment/

[45] http://bigthink.com/daylight-atheism/skepticon-v-impressions

[46] http://scienceblogs.com/denialism/2012/12/05/rebecca-watsons-skepticon-talk-is-not-an-example-of-science-denialism/#comment-22781

criticism at the whole field of evolutionary psychology.[47] — Ken Perrott

By the end of it I did get the impression she was trashing EP overall and not just pop-EP, and it seems I am not the only one.[48] –Peter Ferguson

Ms. Watson is now claiming she wasn't trashing evolutionary psychology as a whole, but just the pop aspects. This is not at all clear from the speech she gave and appears contradictory to her comments in the interview above, where she compares evolutionary psychology to Social Darwinism and eugenics, aptly demonstrating that she can't tell the difference between descriptive and prescriptive claims.[49] –Maria Maltseva

Imagine at the end of the talk, the speaker declared that "mocking neuroscientists never gets old," and that when questioned as to whether there's any good neuroscience, the speaker said "probably?" but then claimed it would have to be totally boring and mainly involve a lot of saying "I don't know."

Would neuroscientists be pissed about such a talk? You betcha. Nobody would buy the "oh, she was just talking about pop neuroscience" defense, or indeed similar arguments about any other branch of science. But this isn't

---

[47] http://openparachute.wordpress.com/2012/12/05/sceptical-arrogance-and-evolutionary-psychology/

[48] http://www.humanisticus.com/2012/12/why-defence-of-rebecca-watsons-ep-talk.html

[49] http://www.skepticink.com/skepticallyleft/2012/12/04/ftb-blogger-stephanie-zvan-makes-a-small-mistake/

the first time I've seen that defense made about sloppy criticisms of evolutionary psychology.[50] –Chris Hallquist

It seems that many people got it "wrong" and in precisely the same way, both before I had said anything, and after. In fact, Watson is not as terrible a communicator as has been suggested. She communicated her opinion about evolutionary psychology quite clearly, if clumsily. Here are broad, sweeping comments from Watson on the subject. Note that she is clear that the source is the field and its researchers, not journalists or the media.

> It's **yet another** study ... and it's not the first time that an evolutionary psychologist has tried to support a shitty stereotype about women (in relation to two studies and allusion to more, emphasis mine, 34:10).

> Evolutionary psychology theories tend to be unfalsifiable (13:17).

> Evolutionary psychologists want to ignore [recent history] and pretend that women's place is in the home (36:36).

> Here are a few resources for you, if you're interested in this sort of stuff, particularly in mocking evolutionary psychologists, which never gets old (45:25).

> The biggest problems with the study, though, are the same problems which are leveled against **evolutionary psychology as a whole** (emphasis mine, 12:25).

---

[50] http://www.patheos.com/blogs/hallq/2012/12/rebecca-watsons-awful-skepticon-5-talk-on-evolutionary-psychology/

Evolutionary psychology requires that our brains evolved twelve thousand to one million years ago and haven't changed since (12:46).

According to the evolutionary psychologists, brains stopped evolving. ... Evolutionary psychology theories tend to be unfalsifiable. A lot of times they'll say these behaviors are written into our genes but they never actually tell us which genes (13:11).

...these examples seem to be completely ignored when evolutionary psychology proponents use present-day groups as proxies (14:58). [Note in context, "proponent" must mean researcher. No one else can be said to "use" a methodological research proxy]

The following remarks are from an interview with Watson shortly after she had given the same talk in Berlin.[51]

Watson: Basically I used my talk as an opportunity to slam evolutionary psychology for half an hour. [...]
Interviewer: Can you tell us a little bit more about evolutionary psychology, what the field actually is?
Watson: Yeah. Well, evolutionary psychology is the idea that humans evolved during the Pleistocene epoch, which we did. But also that our brains evolved, which they did. But that our brains stopped evolving then, so that we currently have Pleistocene brains inside modern bodies. And they ... This isn't necessarily supported by any evidence. [...]
Interviewer: Also, it seems that they are basically guessing a lot.
Watson: Yeah, a lot of it does seem to be. Like one of the examples I give in my talk is this "women evolved to shop"

---

[51] http://soundcloud.com/techniskeptic/rebecca-watson

idea, in which a researcher says that women gathered while men hunted, so shopping is like gathering and visiting cultural institutions is like hunting. And so he relates all those things together without actually giving any more thought or evidence.

Note that Watson's inclusion of Pleistocene centrality as part of the *definition* of evolutionary psychology, according to her, cannot be supported by the evidence. She is not speaking of pop science here, but about the very definition of the field, in her own words. Next, Watson agrees that "they," meaning "evolutionary psychologists" are "guessing a lot." She then gives as an example the first example from her own talk, demonstrating incontrovertibly that, to Watson, there is no distinction between the titular topic of her presentation and general evolutionary psychology. She uses them interchangeably. A moment later, Watson explains that evolutionary psychology was created by bigots and racists to support their bigotry and racism (emphasis mine).

I think there are people who hold misogynist, racist, bigoted ideas, but they value science, and so they will seek out what they consider science in order to support their prejudice. And it's been happening since the beginning of time. I mentioned during my Q&A that evolutionary psychology is not a new thing. It's becoming more and more popular in the last few years, but it's actually evolved from other things, like Social Darwinism, which, you know, got into a lot of trouble over eugenics and things like that. You know, so they change names and they slightly change their viewpoints. So, it's not a new thing, and I do think [evolutionary psychology is] a result of people simply trying to use science to call their prejudice natural.

... **once you look past the headlines and actually look at the studies, what you see over and over and over** again is pseudoscience being passed off as science. You know, they have tons of assumptions that they don't support with the evidence, and they make up Just-So stories that seem to fit the facts. And it only ends up reinforcing stereotypes, which does harm to all of us.

Watson needed three *over*s to drive the point home: the studies are pseudoscientific. Not the headlines. One has to wonder when Watson "looked past the headlines" since her sources seem to have been newspaper clippings, books, and abstracts. Twice, when asked about "good" or "evidence-based" evolutionary psychology, Watson gave a shrugging speculative reply; in Berlin it was "I'm sure there are..." (but you don't know?) and in Missouri it was "prooooobably?" Watson is confident in calling the field pseudoscience rooted in racism and not "necessarily in evidence," but isn't sure if there is any "good" science there or not. In the case of both the interview and the Skepticon Q&A, she makes it clear that a good researcher could only produce dismissals of the field and negative findings:

In Berlin: ... there must be evolutionary psychologists out there who are very careful with their work and who don't make large pronouncements like one I mentioned in my talk ... I'm sure there are researchers who come to a conclusion more like "It's inconclusive whether such and such occurred."

At Skepticon: ... Good evolutionary psychology would be more like, "well, we don't really know what happened in the Pleistocene and we have no evidence for this but maybe this. It's not the sort of thing that makes headlines.

For what sort of topic could it be true that "good" or "evidence-based" investigation will always produce nothing? Creationism, phrenology, and mental telepathy come to mind. In short, it could only be true for a topic you think is nonsense, in which there is no truth to discover. If all that is not enough, at the end of her talk, Watson directed her audience to learn more from people who are explicit and unashamed deniers of evolutionary psychology. Watson recommended Greg Laden, who once wrote[52]:

> One could think of Evolutionary Psychology as the deformed misguided freakish evil sibling of behavioral biology that should have been smothered at birth. Not that I have strong feelings about it or anything.

Amanda Marcotte, also recommended by Watson, tweeted[53]:

> Skepticism of evo psych isn't "anti-science" anymore [sic] than skepticism of phrenology is. Meh.

And as mentioned in the "fake experts" section, Marcotte authored such gems as:

> "Women are born whores," is quite possibly the favorite sacred belief of evolutionary psychology.

Shall we believe that Watson did not intend for her audience to get the idea that general evolutionary psychology is empty and wrong, by telling her audience to go listen to people who say that it is?

---

[52] http://gregladen.com/blog/2011/11/a-foundation-in-mainly-human-behavioral-biology/
[53] https://twitter.com/AmandaMarcotte/statuses/294252921292472320

Questions about Watson's Response

Perhaps I should say "lack of response." One vague paragraph aside, Watson has never responded to my criticism. She did not feel it necessary to answer the list of misrepresentations and errors (at the time of original writing there were twenty-five listed). She accepted a total of three corrections. She never addressed the bulk of these, never answered the "lingering questions" I specifically posed, and never refuted the five tactics of denialists which I established that she used. If I am wrong about all of these, then I would like to correct the account, especially because, as of this writing, an estimated fifty thousand people have read it. If I am right about some or most of it, then Watson should accept the criticism and apologize to Skepticon and her fans for misleading them so sharply. A self-respecting professional skeptic should always concede to the evidence and admit to their own mistakes. I can understand that criticism can be difficult to respond to. But that is what you sign up for when you elect to be a vocal critic yourself.

Conclusion

The main value in this essay is, ideally, not in what it has to say merely about one person's mistakes or even about evolutionary psychology. Rather, it is one story about how the hard problem of skepticism I discussed in the introduction takes form, even in the seemingly unlikeliest of places. This problem is continuing even now. Rebecca Watson is scheduled to speak at Skepticon once again and, at the time of this writing, has been given a blog at Popular Science.[54] She continues to co-host the Skeptics Guide to the Universe podcast,[55] aimed at communicating science to the public.

Scientific skepticism is part of the daily life of an academic researcher. I coauthored a critical review of a mainstream hypothesis. I formally criticized a theory in evolutionary psychology that has stood for years. I have served as a peer-reviewer for several journals, penning as effective a critical appraisal as I could to the research of my colleagues in the field. I do these things in part because I love evolutionary psychology. I know that it is a good science and that a good science gets better with robust criticism. I am excited to be able to play a small part in that, if I can. My last publication in particular was also an exercise in skepticism toward something that I cared about. We need to engage in this kind of skepticism because as we try to figure out how the world works and how it got to be the way that it is; commitments to ego and ideology tend to get in the way. Rebecca Watson is a case study in how easily everything can go wrong, even among the smart and literate people so enamored with skepticism that they've named themselves after it. Skepticism should not be a cudgel for bashing ideas and people we may not like. It should be a lantern that helps us to find our way and perhaps more importantly, to see ourselves more clearly.

---

[54] http://www.popsci.com/category/popsci-authors/rebecca-watson
[55] http://www.theskepticsguide.org/about/rebecca-watson

Acknowledgements

My kind thanks to contributions from Dave Allen and helpful feedback from Daniel Fessler, Martie Haselton, and Robert Kurzban.

References

Adams, C. (2008). The Straight Dope: Was pink originally the color for boys and blue for girls? *StraightDope.com.* Retrieved October 12, 2013, from http://www.straightdope.com/columns/read/2831/ was-pink-originally-the-color-for-boys-and-blue-for-girls

Alcock, J. (2000). Misbehavior How Stephen Jay Gould is wrong about evolution. *Boston Review.* Retrieved May 28, 2013, from http://www.bostonreview.net/BR25.2/alcock.html

Alvergene et al. (2011). Kanazawa's bad science does not represent evolutionary psychology. *Evolutionary Psychology.* Retrieved May 28, 2013, from http://www.epjournal.net/wp-content/uploads/kanazawa-statement.pdf

Alvergne, A., and Lummaa, V. (2010). Does the contraceptive pill alter mate choice in humans? *Trends in Ecology and Evolution, 25*(3), 171–179. Retrieved from http://www.sciencedirect.com/science/article/pii/S 0169534709002638

Apter, T. (2010). Delusions of Gender: The Real Science Behind Sex Differences by Cordelia Fine. *TheGuardian website.* Retrieved May 28, 2013, from http://www.guardian.co.uk/books/2010/oct/11/del usions-gender-sex-cordelia-fine

Armstrong, E. A., England, P., and Fogarty, A. C. K. (2012). Accounting for Women's Orgasm and Sexual Enjoyment in College Hookups and Relationships. *American Sociological Review, 77*(3), 435–462. doi:10.1177/0003122412445802

Arnquist, S. (2009). Testing Evolution's Role in Finding a Mate—NYTimes.com. *The New York Times website.* Retrieved May 28, 2013, from

http://www.nytimes.com/2009/07/07/health/07da
ting.html?_r=0

Barnett, L. (2009). Primitive lot, these shoppers. *Express.co.uk.* Retrieved September 29, 2013, from http://www.express.co.uk/news/weird/86334/Primi tive-lot-these-shoppers

BBC news. (2011). LSE lecturer Dr Satoshi Kanazawa tells of race blog "regret." *BBS News London.* Retrieved May 28, 2013, from http://www.bbc.co.uk/news/uk-england-london-14945110

BBC News. (2011). LSE investigates lecturer's blog over race row. *BBC News.* Retrieved May 28, 2013, from http://www.bbc.co.uk/news/uk-13452699

Bouton, K. (2010). "Delusions of Gender" Peels Away Popular Theories. *The New York Times website.* Retrieved May 28, 2013, from http://www.nytimes.com/2010/08/24/science/24sc ibks.html

Buss, David M., and Meston, C. M. (2009). Why do women have sex? (David Buss and Cindy Meston at CASW 2009). *YouTube.com.* Retrieved May 28, 2013, from http://www.youtube.com/watch?v=KA0sqg3EHm8&f eature=youtu.be&t=5m10s

Buss, David Michael, and Schmitt, D. P. (2011). Evolutionary Psychology and Feminism. *Sex Roles, 64*(9-10), 768–787. doi:10.1007/s11199-011-9987-3

Castleman, M. (2010). Why Do People Have Sex? | Psychology Today. *Psychology Today.* Retrieved May 28, 2013, from http://www.psychologytoday.com/blog/all-about-sex/201011/why-do-people-have-sex

Clark, R., and Hatfield, E. (1989). Gender Differences in Receptivity to Sexual Offers. *Journal of Psychology and Human Sexuality, 2*(1), 39–55. doi:10.1300/J056v02n01_04

Clint, E. K., Sober, E., Garland Jr., T., and Rhodes, J. S. (2012). Male Superiority in Spatial Navigation: Adaptation or Side Effect? *The Quarterly Review of Biology, 87*(4), 289–313. doi:10.1086/668168

Confer, J. C., Easton, J. A., Fleischman, D. S., Goetz, C. D., Lewis, D. M. G., Perilloux, C., and Buss, D. M. (2010). Evolutionary psychology. Controversies, questions, prospects, and limitations. *The American psychologist, 65*(2), 110–26. doi:10.1037/a0018413

Conley, T. D. (2011). Perceived proposer personality characteristics and gender differences in acceptance of casual sex offers. *Journal of Personality and Social Psychology, 100*(2), 309–329. doi:10.1037/a0022152

Cosmides, L., and Tooby, J. (1997). Evolutionary Psychology: A Primer. *Center for Evolutionary Psychology website.* Retrieved September 29, 2013, from http://www.cep.ucsb.edu/primer.html

Enserink, M. (2010a). Elsevier to Editor: Change Controversial Journal or Resign. *Science Insider.* Retrieved September 29, 2013, from http://news.sciencemag.org/2010/03/elsevier-editor-change-controversial-journal-or-resign

Enserink, M. (2010b). Elsevier to Editor: Change Controversial Journal or Resign. *ScienceInsider.* Retrieved May 28, 2013, from http://news.sciencemag.org/scienceinsider/2010/03/elsevier-to-editor-change-contro.html

Fink, B., Hugill, N., and Lange, B. P. (2012). Women's body movements are a potential cue to ovulation. *Personality and Individual Differences, 53*(6), 759–763. Retrieved from http://www.sciencedirect.com/science/article/pii/S0191886912002930

Finkel, E. J., and Eastwick, P. W. (2009). Arbitrary social norms influence sex differences in romantic selectivity. *Psychological science, 20*(10), 1290–5. doi:10.1111/j.1467-9280.2009.02439.x

Fleischman, D. S. (2013). Sentientist. *Sentientist.com.* Retrieved from http://sentientist.org/

Forbes, P. (2013). Paleofantasy: What Evolution Really Tells Us About Sex, Diet, and How We Live by Marlene Zuk—review. *The Guardian.* Retrieved May 28, 2013, from http://www.guardian.co.uk/books/2013/apr/24/paleofantasy-evolution-sex-diet-review

Gangestad, S. W., Thornhill, R., and Garver, C. E. (2002). Changes in women's sexual interests and their partners' mate-retention tactics across the menstrual cycle: evidence for shifting conflicts of interest. *Proceedings. Biological sciences / The Royal Society, 269*(1494), 975–82. doi:10.1098/rspb.2001.1952

Gendered Division Of Labor Gave Modern Humans Advantage Over Neanderthals. (2006). *ScienceDaily.* Retrieved May 28, 2013, from http://www.sciencedaily.com/releases/2006/12/061204123302.htm

George, A. (2013a). Marlene Zuk's Paleofantasy book: Diets and exercise based on ancient humans are a bad idea. *Slate.* Retrieved May 28, 2013, from http://www.slate.com/articles/health_and_science/new_scientist/2013/04/marlene_zuk_s_paleofantasy_book_diets_and_exercise_based_on_ancient_humans.html

George, A. (2013b). Should we aim to live like cavemen? *NewScientist.* Retrieved May 28, 2013, from http://www.newscientist.com/article/mg21729090.400-should-we-aim-to-live-like-cavemen.html

Goldacre, B. (2007). Imaginary numbers—Bad Science. *Bad Science.* Retrieved May 28, 2013, from http://www.badscience.net/2007/09/imaginary-numbers/

Gonzaga, G. C., Haselton, M. G., Smurda, J., Davies, M. sian, and Poore, J. C. (2008). Love, desire, and the suppression of thoughts of romantic alternatives☆. *Evolution and Human Behavior*, *29*(2), 119–126. doi:10.1016/j.evolhumbehav.2007.11.003

Grammer, K., Renninger, L., and Fischer, B. (2004). Disco clothing, female sexual motivation, and relationship status: is she dressed to impress? *Journal of sex research*, *41*(1), 66–74. doi:10.1080/00224490409552214

Guéguen, N. (2011). Effects of solicitor sex and attractiveness on receptivity to sexual offers: a field study. *Archives of sexual behavior*, *40*(5), 915–9. doi:10.1007/s10508-011-9750-4

Guide for authors. (2012). *Medical Hypotheses*. Retrieved May 28, 2013, from http://www.elsevier.com/journals/medical-hypotheses/0306-9877/guide-for-authors

Hagen, E. (2004). Evolutionary Psychology FAQ. *University of California Santa Barbara Anthropology website*. Retrieved May 28, 2013, from http://www.anth.ucsb.edu/projects/human/epfaq/e vpsychfaq_full.html#holocene

Halpern, D. F. (2010). How Neuromythologies Support Sex Role Stereotypes. *Science*, *330*(6009), 1320–1321. doi:10.1126/science.1198057

Haselton, M. G., and Gildersleeve, K. (2011). Can Men Detect Ovulation? *Current Directions in Psychological Science*, *20*(2), 87–92. doi:10.1177/0963721411402668

Haselton, Martie G, and Gangestad, S. W. (2006). Conditional expression of women's desires and men's mate guarding across the ovulatory cycle. *Hormones and behavior*, *49*(4), 509–18. doi:10.1016/j.yhbeh.2005.10.006

Haselton, Martie G, Mortezaie, M., Pillsworth, E. G., Bleske-Rechek, A., and Frederick, D. A. (2007). Ovulatory shifts in human female ornamentation: near ovulation, women dress to impress. *Hormones and behavior, 51*(1), 40–5. doi:10.1016/j.yhbeh.2006.07.007

Havlicek, J., Dvorakova, R., Bartos, L., and Flegr, J. (2006). Non-Advertized does not Mean Concealed: Body Odour Changes across the Human Menstrual Cycle. *Ethology, 112*(1), 81–90. doi:10.1111/j.1439-0310.2006.01125.x

Holmes, D. (2012). MMU | Research Institute for Health and Social Change. *Manchester Metropolitan University website*. Retrieved May 28, 2013, from http://www.rihsc.mmu.ac.uk/staff/profile.php?surname=Holmes&name=David

Hurlbert, A. C., and Ling, Y. (2007). Biological components of sex differences in color preference. *Current biology : CB, 17*(16), R623–5. doi:10.1016/j.cub.2007.06.022

Kalant, H., Pinker, S., Kalow, W., and Gould, S. J. (1997). Evolutionary psychology: An exchange. *New York Review of Books*, (44), 55–56.

Kaplan, M. (2009, June 2). Role reversal undermines speed-dating theories. *Nature*. Nature Publishing Group. doi:10.1038/news.2009.537

Khamsi, R. (2007). Women may be hardwired to prefer pink. *NewScientist.com*. Retrieved from http://www.newscientist.com/article/dn12512-women-may-be-hardwired-to-prefer-pink.html#.UlfRN1Cfh1F

Kirchengast, S., and Gartner, M. (2002). Changes in fat distribution (WHR) and body weight across the menstrual cycle. *Collegium antropologicum, 26 Suppl*, 47–57. Retrieved from http://www.ncbi.nlm.nih.gov/pubmed/12674835

Krug, R., Mölle, M., Fehm, H. L., and Born, J. (1999). Variations across the menstrual cycle in EEG activity during thinking and mental relaxation. *Journal of Psychophysiology*, (13), 163–172.

Kruger, D., and Byker, D. (2009). EVOLVED FORAGING PSYCHOLOGY UNDERLIES SEX DIFFERENCES IN SHOPING EXPERIENCES AND BEHAVIORS. *Journal of Social, Evolutionary, and Cultural Psychology*, *3*(4), 328–342.

Kuhn, S. L., and Stiner, M. C. (2006). What's a mother to do? A hypothesis about the division of labor and modern human origins. *Current Anthropology*. Retrieved September 29, 2013, from http://www.academia.edu/968719/Whats_a_mother_to_do_A_hypothesis_about_the_division_of_labor_and_modern_human_origins

Kurzban, R. (2013a). Mind Design | Psychology Today. *Psychology Today*. Retrieved May 28, 2013, from http://www.psychologytoday.com/blog/mind-design

Kurzban, R. (2013b). Evolutionary Psychology blog. *Evolutionary Psychology*. Retrieved from http://www.epjournal.net/blog/

Kuukasjarvi, S. (2004). Attractiveness of women's body odors over the menstrual cycle: the role of oral contraceptives and receiver sex. *Behavioral Ecology*, *15*(4), 579–584. doi:10.1093/beheco/arh050

Leach, B. (2009). Shopping is "throwback to days of cavewomen." *The Telegraph*. Retrieved May 27, 2013, from http://www.telegraph.co.uk/news/newstopics/howaboutthat/4803286/Shopping-is-throwback-to-days-of-cavewomen.html

Lombardo, W. K., Cretser, G. A., and Roesch, S. C. (2002). For Crying Out Loud — The Differences Persist into the '90s, *45*(October 2001), 529–547.

Lovgren, S. (2006). Sex-Based Roles Gave Modern Humans an Edge, Study Says. *National Geographic website*. Retrieved May 28, 2013, from http://news.nationalgeographic.com/news/2006/12/061207-sex-humans.html

Lucas, M. (2012). Cordelia Fine, Delusions of Gender: How our Minds, Society, and Neurosexism Create Difference. *Society, 49*(2), 199–202. doi:10.1007/s12115-011-9527-3

Manning, J. T., Scutt, D., Whitehouse, G. H., Leinster, S. J., and Walton, J. M. (1996). Asymmetry and the menstrual cycle in women. *Ethology and Sociobiology*, (17), 129–143. Retrieved from http://www.journals.elsevierhealth.com/periodicals/ensold/article/0162-3095(96)00001-5/references

McDonald, M. M., Asher, B. D., Kerr, N. L., and Navarrete, C. D. (2011). Fertility and intergroup bias in racial and minimal-group contexts: evidence for shared architecture. *Psychological science, 22*(7), 860–5. doi:10.1177/0956797611410985

Meston, C. M., and Buss, D. M. (2007). Why humans have sex. *Archives of sexual behavior, 36*(4), 477–507. doi:10.1007/s10508-007-9175-2

Miller, G., Tybur, J., and Jordan, B. (2007). Ovulatory cycle effects on tip earnings by lap dancers: economic evidence for human estrus? *Evolution and Human Behavior, 28*(6), 375–381. doi:10.1016/j.evolhumbehav.2007.06.002

Miller, L. (2013). "Paleofantasy": Stone Age delusions. *Salon.com*. Retrieved from http://www.salon.com/2013/03/10/paleofantasy_stone_age_delusions/

Miller, S., and Maner, J. (2010). Evolution and relationship maintenance: Fertility cues lead committed men to devalue relationship alternatives.

Journal of Experimental Social Psychology, 46(6), 1081–1084. doi:10.1016/j.jesp.2010.07.004

Navarrete, C. D., Fessler, D. M. T., Fleischman, D. S., and Geyer, J. (2009). Race bias tracks conception risk across the menstrual cycle. *Psychological science, 20*(6), 661–5. doi:10.1111/j.1467-9280.2009.02352.x

Navarrete, C. D., McDonald, M. M., Mott, M. L., Cesario, J., and Sapolsky, R. (2010). Fertility and race perception predict voter preference for Barack Obama. *Evolution and Human Behavior, 31*(6), 394–399. doi:10.1016/j.evolhumbehav.2010.05.002

Nuwer, R. (2013). Men Are Better Navigators Than Women, But Not Because of Evolution. *The Smithsonian.com.* Retrieved May 28, 2013, from http://blogs.smithsonianmag.com/smartnews/2013/02/men-are-better-navigators-than-women-but-not-because-of-evolution/

Penke, L., Borsboom, D., Johnson, W., Kievit, R. A., Ploeger, A., and Wicherts, J. M. (2011). Evolutionary Psychology and Intelligence Research Cannot Be Integrated the Way Kanazawa (2010) Suggested. *American Psychology, 66*(9), 916–7. Retrieved from http://www.larspenke.eu/pdfs/Penke_et_al_in_press_-_Kanazawa_commentary.pdf

Pillsworth, E. G., and Haselton, M. G. (2006). Male sexual attractiveness predicts differential ovulatory shifts in female extra-pair attraction and male mate retention. *Evolution and Human Behavior, 27*(4), 247–258. doi:10.1016/j.evolhumbehav.2005.10.002

Pinker, S. (1997). *How the Mind Works* (p. 660). New York: W W Norton and Co Inc. Retrieved from http://www.amazon.com/How-Mind-Works-Steven-Pinker/dp/0393045358

Pinker, S. (2012). Steven Pinker's correspondence with journalist Dan Slater, 2012. Retrieved May 28, 2013,

from
http://stevenpinker.com/files/pinker/files/pinker_c
orredspondence_with_dan_slater.pdf

Provost, M. P., Quinsey, V. L., and Troje, N. F. (2008).
Differences in gait across the menstrual cycle and
their attractiveness to men. *Archives of sexual
behavior*, *37*(4), 598–604. doi:10.1007/s10508-007-
9219-7

Roberts, S. C., Havlicek, J., Flegr, J., Hruskova, M., Little,
A. C., Jones, B. C., … Petrie, M. (2004). Female facial
attractiveness increases during the fertile phase of
the menstrual cycle. *Proceedings. Biological sciences
/ The Royal Society*, *271 Suppl* (Suppl_5), S270–2.
doi:10.1098/rsbl.2004.0174

SD-agencies. (2009). Shopping is "throwback to days of
cave women:" study. *Shenzhen Daily*. Retrieved
September 29, 2013, from
http://szdaily.sznews.com/html/2009-
02/26/content_528445.htm#

Self Magazine. (2011). TresSugar/SELF Magazine Casual
Sex Survey. *Self Magazine*. Retrieved September 30,
2013, from
http://www.tressugar.com/TresSugarSELF-
Magazine-Casual-Sex-Survey-Results-
16742092?image_nid=16742092

Singh, D., and Bronstad, P. M. (2001). Female body odour
is a potential cue to ovulation. *Proceedings. Biological
sciences / The Royal Society*, *268*(1469), 797–801.
doi:10.1098/rspb.2001.1589

Slater, D. (2013). Darwin Was Wrong About Dating. *New
York Times website*. Retrieved May 28, 2013, from
http://www.nytimes.com/2013/01/13/opinion/sun
day/darwin-was-wrong-about-
dating.html?pagewanted=all

Switek, B. (2010). The earliest known pelican reveals 30
million years of evolutionary stasis in beak

morphology. *Wired.com, 152*(1), 15–20.
doi:10.1007/s10336-010-0537-5

Symonds, C. S., Gallagher, P., Thompson, J. M., and
Young, A. H. (2004). Effects of the menstrual cycle on
mood, neurocognitive and neuroendocrine function
in healthy premenopausal women. *Psychological
medicine, 34*(1), 93–102. Retrieved from
http://www.ncbi.nlm.nih.gov/pubmed/14971630

Symons, D. (2000). Regarding V.S. Ramachandran. *Center
for Evolutionary Psychology website*. Retrieved May
28, 2013, from
http://www.cep.ucsb.edu/ramachandran.html

Telegraph Media Group. (2007). Jessica Alba has the
perfect wiggle, study says. *Telegraph*. Retrieved May
28, 2013, from
http://www.telegraph.co.uk/news/uknews/1561306
/Jessica-Alba-has-the-perfect-wiggle-study-
says.html

Thornhill, R., and Palmer, C. T. (2001). *A Natural History
of Rape: Biological Bases of Sexual Coercion* (p. 272).
MIT Press. Retrieved from
http://books.google.com/books?hl=en&lr=&id=xH6v
-nB6EegC&pgis=1

Tooby, J., and Cosmides, L. (2000). Response to Coyne.
Center for Evolutionary Psychology website. Retrieved
May 28, 2013, from
http://www.cep.ucsb.edu/tnr.html

Van Hemert, D. A., van de Vijver, F. J. R., and
Vingerhoets, A. J. J. M. (2011). Culture and Crying:
Prevalences and Gender Differences. *Cross-Cultural
Research, 45*(4), 399–431.
doi:10.1177/1069397111404519

Villotte, S., Churchill, S. E., Dutour, O. J., and Henry-
Gambier, D. (2010). Subsistence activities and the
sexual division of labor in the European Upper
Paleolithic and Mesolithic: evidence from upper limb

enthesopathies. *Journal of Human Evolution, 59*(1), 35–43. doi:10.1016/j.jhevol.2010.02.001

Ward, A. F. (2012). Scientists Probe Human Nature--and Discover We Are Good, After All. *ScientificAmerican.com.* Retrieved May 28, 2013, from http://www.scientificamerican.com/article.cfm?id=scientists-probe-human-nature-and-discover-we-are-good-after-all

Wayman, E. (2012). When Did the Human Mind Evolve to What It is Today? *The Smithsonian.* Retrieved May 28, 2013, from http://www.smithsonianmag.com/science-nature/When-Did-the-Human-Mind-Evolve-to-What-It-is-Today-160374925.html

Weierstall, R., Schauer, M., and Elbert, T. (2013). Psychology of War Helps to Explain Atrocities. *Scientific American.* Retrieved May 28, 2013, from http://www.scientificamerican.com/article.cfm?id=psychology-war-helps-explain-atrocities

Wilson, D. S. (2013). Evolution: This View of Life Magazine. *Evolution: This View of Life.* Retrieved from http://www.thisviewoflife.com/index.php

Zimmer, C. (2013). Of Men, Navigation, and Zits— Phenomena. *National Geographic website.* Retrieved May 28, 2013, from http://phenomena.nationalgeographic.com/2012/12/28/of-men-navigation-and-zits/

Zuk, M. (2013). *Paleofantasy: What Evolution Really Tells Us about Sex, Diet, and How We Live.* W. W. Norton and Company. Retrieved from http://www.amazon.com/Paleofantasy-Evolution-Really-Tells-ebook/dp/B007Q6XM1A

Appendix
Science Denialism at a Skeptic Conference

90 Self-contradictions, errors, false and misleading claims, and misrepresentations made by Rebecca Watson

This list shows a consistent pattern of confirmation bias, cherry-picking, intellectual dishonesty, self-contradiction, and pretension to unowned scientific literacy. Some items are minor and may appear nit-picky. They are only included because they are numerous and help illustrate a lackadaisical attitude and a consistent pattern of poor research. Video indices referring to the YouTube video of her talk precede each quotation or paraphrasing of her statements.

 Watson said that she knows about the "scientific fact" that "girls evolved to shop" because "this is a science story that has appeared in the science section of major newspapers around the world not once, but **several times**. Here's the first time I noticed it, this is in February 2009" (emphasis mine). An image of a *Telegraph* web article titled *Shopping is 'throwback to days of cavewomen'* appeared on the screen.

1

The article has never appeared in the science section of any major newspaper. It has appeared in *The Telegraph* and the tabloid *The Daily Express*. *The Telegraph* and *Daily Express* filed the story under the sections "Weird" and not "Science" (Barnett, 2009; Leach, 2009). If the story ever appeared in any "major newspaper" after those, I was unable to find it. It was also run in an English-language Chinese newspaper called the Shenzhen Daily, which also filed it as weird (SD-agencies, 2009).

2

In a talk alleging sexism of others, Watson has curiously chosen to refer to all female persons as "girls" in the title. No article or paper mentioned in her talk uses such sexist language. Watson never explained the usage.

`1:50` Watson said the *Telegraph* article "describes a "study" [Watson made the air quotes gesture] done by Dr. David Holmes of Manchester Metropolitan University who said that women love to shop."

3

Dr. Holmes is never quoted as saying "women love to shop" or anything near to it. He only used gender-neutral terms. The headline reading "cavewomen" was an editorial choice of the publication, not the assertion of Dr. Holmes, as Watson falsely indicated (Leach, 2009).

`1:52` *Dr. David Holmes... who said that women love to shop because, and I quote, "skills that were learned as cavemen and women were now being used in shops."*

4

This is not a quote of Dr. Holmes, but Ben Leach paraphrasing Holmes, in spite of Watson's misleading use of the phrase "and I quote" (Leach, 2009). The *Express*, which clearly covered the same story on the same day from the same source, probably parsed Holmes more accurately: *Research suggests modern-day Wilma Flintstones are only using the instincts they inherited from their hunter—gatherer forebears.* Note, "instincts" and "inherited," not "skills" and "learnt" (Barnett, 2009).

`2:30` *Now I'm no scientist like Dr. Holmes but I found a few problems with his line of reasoning. For instance you don't generally inherit traits that are learned behaviors. For instance my father is very good at playing the drums, I cannot play the drums. It's weird that I wasn't born playing the drums.*

5

Dr. Holmes never said any such thing. Watson is responding to the journalist's poor phrasing and misattributing it to Dr. Holmes (Leach, 2009).

`3:26` *Number three, you don't gather in the cave. If you only gather in the cave all you eat is stalactite mushroom soup, you have to leave the cave to gather things.*

6

This reasoning was falsely attributed to Dr. Holmes, it was not in the article. Caves offered some hominins shelter from the elements, not foraging venues. Watson appeared to misconstrue the phrase "gathered in caves with fires at the entrance" (Leach, 2009).

3:55 *So if we actually inherited that learned behavior of leaving the cave to shop this is what our shopping malls would look like.* The overhead projection then showed a cartoon of modern consumer products placed on bushes.

7

"Inherited learned behavior" is a paraphrase of a nonsensical line written by Ben Leach, not by Dr. Holmes (Leach, 2009). As noted, the *Express* (which the *Telegraph* cites as a source) used the language "inherited" not "learnt."

8

Watson appeared to assert here that heritable psychological traits could only be evidenced if confined to the most superficial features. This is a naive and unscientific understanding of biology. A Bowerbird may construct a nest from plastic bottle caps. This is not because it evolved to locate and use plastic, but because it evolved to produce elaborate and colorful nests and plastic bottle caps that humans happened to have left about are one way to accomplish that for some modern Bowerbirds. Similarly, humans evolved to forage to locate sustenance and to make use of any available shelter against the elements, not to idiotically home in on bushes or caves.

4:18 *Problem number four. If women have been the ones who have been most interested in fashion since the Pleistocene was King Louis XIV some fabulous outlier?* [Watson showed a painting of an elaborately costumed monarch].

9

Watson responded to a claim that does not appear in the article. It does not mention fashion or clothing at all. We know she was not speaking more broadly because of the words "Problem number four..." a continuation of a series she began (see 2:26) with "I found a few problems with [Dr. Holmes's] line of reasoning." King Louis XIV is utterly irrelevant to this section of the talk about shopping as a 17th century monarch would not have done much of what we call shopping. His clothing and personal effects would have been custom-produced for him.

10
Watson incorrectly referred to Louis XVI as Louis XIV.

11
Watson appeared not to understand the scientifically important concept "outlier" as she failed to recognize that the King of France would count as an outlier in most respects, certainly in terms of fashion.

`4:33` *In the end though this doesn't matter because this isn't actually science (surprise!). The end of the article did helpfully explain "the study was commissioned by Manchester Arndale Shopping Centre in a response to a rise in January visitors." All of the best studies I find are commissioned by shopping centers. This is actually marketing disguised as science.*

12
Funding source alone does not necessarily discredit any particular study. We do not know if Dr. Holmes study is science because it was not published and has not been read by Watson. The *Telegraph* article may well be described as marketing, but such cannot merely be assumed about a paper that one has not even read. Surely that is necessary before declaring something as not science.

Twenty-three minutes later Watson cited a dating survey commissioned and produced by the Cosmo-esque *Self* magazine as evidence about female sexuality as a means of refuting published scientific research (See point 55).

13

The selfish interest of the Arndale Shopping Centre is to get useful, reliable information. According to the *Express*, the study was commissioned to determine possible reasons why the center was having record sales during a down economy. Its business manager was quoted as saying, *It seems our gatherer instincts are coming to the fore and affecting the way we shop in these testing times.* Notable is the phrasing *the way we shop*, not *why women shop*. This is just the sort of concern expected from a business owner who wants to make their store more appealing to customers (Barnett, 2009).

14

Through the above misattributions, Watson paints Dr. Holmes as complicit in sexism, if not outright sexist in his work. Dr. Holmes's other work includes fourteen presentations at professional conferences on the topics of stalking, domestic violence, and rape. Many of these appear to have been forensic talks designed to help arrest criminal offenders. This may prove nothing definitively, but his presumptive likelihood of sexist attitudes is dubious. Combined with the lack of any evidence for the contrary, Watson's implications about Holmes are irresponsible (Holmes, 2012).

6:15 Watson paraphrased Ben Goldacre's account of a public relations company (Clarion Communications) soliciting a scientist to do a study they wished to show a preset result:

We're conducting a survey into the celebrity top ten sexiest walks.... We would like help from a doctor of psychology, or someone similar, who could come up with equations to back up our findings as we feel that having an expert comment on our equation will give our story more weight. We haven't done the survey yet but we know what results we want to achieve. We want Beyoncé to come out on top followed by other celebrities with curvy legs.

At index 7:13 Watson continued, *So you might think like nooo scientist would fall for this; no scientist with an ounce of morality would fall for this. But somebody did.* Watson then showed a slide featuring a *Telegraph* article (Telegraph Media Group, 2007) *"Jessica Alba has the perfect wiggle, study says,"* and continued,

The scientist that they ended up quoting was angry because they attributed his research to a whole team at Cambridge and they just ignored who he actually said had the sexiest wiggle.

15
According to Ben Goldacre, the scientist in question is Prof. Richard Weber of Cambridge. Weber was not angry because they misattributed his work to a team, but because Clarion published the entirely inaccurate release *at all*. About it, Goldacre quoted Weber as writing,

The Clarion press release was not approved by me and is factually incorrect and misleading in suggesting there has been any serious attempt to do mathematics here. No such thing has happened. No "team of Cambridge mathematicians" has been involved in producing the results that have been reported. I do not endorse what the press release says. I did not approve it and would not have done so if asked. I have emailed my contact in Clarion Communication to ask for an explanation, but I have had a reply that she is on holiday.

Clarion asked me to help by analyzing survey data on from 800 men in which they were asked to rank 10 celebrities for "sexiness of walk." Jessica Alba was 7th on the list, near the bottom. I reported that there was little one could conclude from the data on the 10 names ... I suggested that as a bit of fun and nonsense, but no more, that they could say something like the following: "I have studied how 10 celebrities have ranked for "sexiness of walk" in relation to their bust-waist-hip measurements.... I fear that the Clarion press release is an example of disingenuous and perverted use [of] this simple remark, not of any bad science. I trust you will not wish to follow their lead.

Weber did no research, and after doing statistical analysis, found nothing to support the conclusion Clarion wanted, and told them so. Clarion then disregarded it all and published apparently fabricated results falsely attributing Cambridge without Weber's consent and against his wishes (Goldacre, 2007).

8:51 *[Evolutionary psychology is] a field of study based on belief that the human brain as it exists today evolved completely during the Pleistocene era when humans lived as hunter-gatherers.*

16

Evolutionary psychologists stipulate that change during the Holocene has occurred; it is merely limited because 11,000 years is a relatively short amount of evolutionary time for a species with a 20-year reproductive cycle.

The study of recent evolution is sometimes avoided for several good reasons. One is that large "big picture" understandings of the evolution of the brain are unanswered, making the asking of many smaller questions impossible.

Also, claims about recent evolution are the kind that have been politically abused by genetic supremacists who wish to claim one "race" is superior.

Our ancestors lived as hunter-gatherers for a thousand times longer than they lived any other way (Cosmides and Tooby, 1997). During the same period, brain size underwent a massive increase. It is not reasonable to imagine this period did not leave lasting marks on our psychology, regardless of any recent (<11,000 years) evolutionary change.

Even if our minds evolved in substantial ways in the last 11,000 years, those changes would be based on the substrate mind forged over the previous million years which would leave lasting marks just the way that whales have hip bones, lungs, and move by vertical undulation of the body like land mammals and not like fish, in spite of tens of millions of years in the water and periods of rapid evolution for an aquatic lifestyle.

8:11, 9:41 Daniel Kruger and Dreyson Byker's 2009 study incorrectly cited as "University of Chicago study."

17

Dr. Kruger is faculty at the University of Michigan. His study is titled "Evolved foraging psychology underlies sex differences in shopping experiences and behaviors" and was published in the *Journal of Social, Evolutionary, and Cultural Psychology* (Kruger and Byker, 2009).

8:11 *[R]esearchers at Chicago also came up with the theory that women evolved to shop, the scientific theory, and I'm using "scientific theory" in the same way as Creationists use scientific theory, which is not scientific theory.*

18

Daniel Kruger and Dreyson Byker's paper does not assert that "women evolved to shop" but a milder thesis (quoting from the abstract) equally about men and women (Kruger and Byker, 2009):

These results suggest that shopping experiences and behaviors are influenced by sexually divergent adaptations for gathering and hunting.

Furthermore, Kruger and Byker demonstrated understanding of the many factors which influence human behavior, including socialization. In the concluding section they wrote,

It must be noted that cultural and social norms likely impact on people's shopping experiences and behaviors, and the authors are not ruling against these influences. For example, with regards to navigation, it is quite possible that girls have fewer opportunities than boys to engage in activities that develop directional skills. Doguand Erkip (2000) propose that women might be encouraged to shop more during their development, and thus, they view stores or malls differently than men and pay particular attention to objects, as stores revolve around the displaying of items.

10:11 *So back in the day men were hunters and women were gatherers, and now men like museums where women prefer shopping because the researcher in question noticed this on a trip to Prague. He went with some friends and all the men in the group wanted to go and see cultural attractions and all the women wanted to go shopping. ... So he is determined that visiting museums is like hunting and shopping is like gathering, ergo, science!*

19

Watson demonstrated an unfamiliarity with how scientists actually produce research. Hypotheses are guesses that scientists make, usually based on their observations or on existing literature but extending into a new area. Social scientists regularly find inspiration in behaviors they observe in their own lives. What makes a guess scientific or not is whether it is amenable to empirical testing, not how plausible *a priori* Watson, or anyone else, finds it.

20

Watson suggested that Kruger was arbitrarily selecting which behavior seemed more like hunting and which more like gathering. This is a confusion about what inspired a topic of study and what the actual research methods were. Kruger and Byker surveyed opinions, attitudes, and strategies related to shopping. The word "museum" does not even appear in the paper and no one was asked what types of sightseeing activities they prefer. Watson's opinion of Kruger's musing in a media piece is not relevant to the validity of the research or findings.

21

Kruger and Byker's paper could only be ruled unscientific by consideration of the theoretical basis and methodologies they used in their study, not by the colorful and manipulated quotes that appear in the media. There was no indication that Watson has read it.

12:45 *Evolutionary psychology requires that our brains evolved twelve thousand to one million years ago and haven't changed since, which doesn't actually fit in with what we understand about evolution. We're not finished evolving.* Watson then gave, as an example of recent evolutionary change, our ability to consume animal milk.

22

Watson appeared to believe that little change over twelve thousand years is incongruent with evolutionary theory. In fact, species sometimes undergo little change over millions of years. Fossil evidence shows that Pelicans, for example, have changed little over the course of thirty million years. This is 2500 times longer than the duration which Watson finds implausible (Switek, 2010).

Evolutionary psychologists state plainly that changes are possible, see point 16 (Hagen, 2004).

23

The ability to drink (animal) milk throughout the lifespan *is* a recent change, namely the mutation causing the enzyme lactase to be produced throughout the lifespan and not merely during infancy. However, this observation does not support Watson's argument because:

- Humans have shown virtually no major physiological changes in the last 10,000 years. Lactase is the exception, not the rule. This is evidence the mind is unlikely to have many large recent changes, not the contrary.

- The mutation in the case of lactase is a simple one, an amendment to an existing enzyme's production schedule. Evolutionary psychologists largely focus on complex behaviors requiring equally complex genetic changes and thus far more time than the one observed with lactase.

- Watson's example relied on many assumptions about the conditions of the Pleistocene. For example, that humans had not yet domesticated cattle to get milk from, and that humans did not normally consume milk beyond childhood. This is inconsistent with her insistence that the Pleistocene is uncertain and unknowable for the purposes of scientific consideration at point 25.

13:27 *A lot of times [evolutionary psychologists] say this stuff, these behaviors are written into our genes but they never actually tell us which genes. There's no evidence to support it.*

24

The implication that gene(s) must be identified before an adaptation is demonstrated is specious. The remark that there is "no evidence to support it" is equally mistaken. To quote Confer et al. 2010,

Adaptations are typically defined by the complexity, economy, and efficiency of their design and their precision in effecting specific functional outcomes, not by the ability of scientists to identify their complex genetic bases (Williams, 1966). For example, the human eye is indisputably an adaptation designed for vision, based on the design features for solving the particular adaptive problems such as detecting motion, edges, colors, and contrasts. The universal and complex design features of the eyes provide abundant evidence that they are adaptations for specific functions, even though scientists currently lack knowledge of the specific genes and gene interactions involved in the visual system (Confer et al., 2010).

The founder of genetics, Gregor Mendel, had no idea what genes actually were, let alone which did what. Charles Darwin had no idea what a gene *was*, as he drafted his theory of evolution based on the observations of apparent adaptations. One does not have to fully understand the mechanism to make valid inferences from observation.

13:39 *[A] problem leveled against evolutionary psychology as a whole: ... It's shocking how little we know about our ancestors. We have some guesses, but the 2 million years that made up that era were incredibly varied in terms of climate and in terms of environment and most likely the lives lead by our Pleistocene ancestors were just as varied... a lot of what we assume about them is taken from present day hunter-gatherer cultures.*

25

Evolutionary psychologists only lean heavily on non-controversial facts about the past. For example, pregnancy involves numerous costs, and we therefore expect that females in many species will be pickier about mating than will males. This prediction has strong empirical support for both humans and other animals (Hagen, 2004).

The ancestral environment is not restricted to the Pleistocene. Powerful hormones such as testosterone, for example, date back half a billion years. Breast feeding originated in early mammals eons before primates existed.

26

Contemporary hunter-gatherer societies are presumed, quite sanely, to be *more like* our ancestors than we are, but not to be exactly like them. Existent small-scale societies are used as a proxy for the past *only* when the feature under consideration is reasonably expected to have been relevant in the ancient past, such as how pre-state peoples might cope with parasites.

13:18 *Evolutionary psychology theories are unfalsifiable* [on slide].

27

At appendix points 29, 50, 55, 58, and video index 24:30, 39:00, Watson cited research which she believed refuted evolutionary psychology hypotheses. Watson declared said hypotheses to be false based on that research, proving conclusively that she knows such hypotheses are falsifiable, despite her claim here.

28

Confer et al. 2010 thoroughly refuted this claim, citing the theories of domain-specific memory and error management which have produced dozens of testable predictions which have been empirically tested.

I authored a paper criticizing one particular evolutionary psychology hypothesis, which is to say, it was a test of that hypothesis (Clint et al. 2012).

`14:17` *There are some contemporary African cultures in which men are the primary gatherers* (posed as objection to the notion of a knowable stereotypic Pleistocene environment).

29

The anthropological record is clear that these cases are the exception, and that these exceptions happen for reasons based in ecology. To quote the study by Dr. Kruger which Watson cited at points 17 and 18:

These are aggregate tendencies, as men sometimes gather (Halpern, 1980) and women sometimes hunt (Noss, 2001). The sex reversal in activities usually take place under special conditions, such as male gathering when meat is scarce during the dry season, and these men often specialized in carrying heavy loads rather than searching for food (Halpern, 1980). In environments where food is more abundant and less

221

seasonal, males gather proportionally more so than in more scarce and seasonal environments (Marlowe, 2007). Women do not hunt as often as men, and usually hunt more reliable small game when caloric return is relatively high compared to gathering alternatives (Noss, 2001). For example, Agta women in central Africa hunt in groups with nets for small game, and do not hunt when they have infants, a limitation that men do not face (Noss, 2001). **It is important to recognize that evolution by selection does not require or imply absolutes; there will often be a few examples that contrast with the general pattern**. *Therefore, in general men tend to hunt and women tend.* [Emphasis mine] (Kruger and Byker, 2009).

`14:48` *Recent research by anthropologist Steven Kuhn suggests that there was no sexual division of labor prior to the upper Paleolithic.*

30

Watson called Steven Kuhn and Mary Stiner's (Watson did not credit Stiner) paper "recent research" even though it was published in 2006, a bit like saying Shakira's "Hips Don't Lie" is a recent hit pop song. This gave the misleading impression the paper was new. In reality, there has been quite a lot of further study and research after it which has largely not supported the hypothesis.

Kuhn's hypothesis is just one of several attempting to explain why *Homo sapiens* out-competed Neanderthals. Viable competing theories include symbol use and the invention of projectile weaponry. Watson may not realize Kuhn was saying the sexual division of labor made *Homo sapiens* better at surviving than the Neanderthals, which is identical to other arguments for adaptive sex differences that

Watson finds so sexist. She made no remark upon it here, when she believed it supported her views.

Further, there is no consensus among archaeologists that the physical evidence proves upper Paleolithic humans were the first with a sexual division of labor, nor that Neanderthal's lacked them.

Subsequent studies of physical evidence have concluded Neanderthal women did not hunt as Kuhn supposed (Villotte et al., 2010).

Watson cited Kuhn as proof that modern hunter-gatherers are not indicative of the past. Kuhn shares this view, but clarifies that such knowledge is useful in formulating models of the past. He wrote: ...*models developed from data on recent hunter-gatherers are most informative precisely when they prove to be inadequate predictors of patterns encountered in the Paleolithic record.* In using the present hunter-gatherer studies to detect disjunctive predictions about the Paleolithic hominins, he is engaging in the same comparative reasoning as the evolutionary psychologists which Watson criticized at points 25 (13:39) and 29 (14:17) (Kuhn and Stiner, 2006).

`15:26` *This is why there are tons of people who, particularly scientists, who think that a lot of the pop evolutionary psychology is "just so stories" as Stephen Jay Gould noted.* (15:26)

31

It is unclear to whom "tons of scientists" refers. The only one named is Stephen Jay Gould who used the term "just so stories," 33 years ago, and which is not considered cogent criticism on the grounds of being trite and glib. Scholarly rebuttals to Gould's criticism have been authored by John Alcock and Steven Pinker (Alcock, 2000; Kalant et al., 1997).

32

Watson mischaracterized Gould as "noting" something about "pop evolutionary psychology." Gould aimed his criticisms at "selectionists" in biology and sociobiology as wholes and not "pop" anything.

`15:41` *The accusation that [many scientists] make is that evolutionary psychology researchers first identify a behavior, like shopping. They assume it has evolved, as a response to environmental pressures, they don't need evidence for that. And then they find anything in our ancient past that might be relevant to that.*

33

No one believes that just any behavior *must* be an adaptation. Behaviors and features are chosen for testing when they show coherent function which is not explained by existing understanding. Even then, a hypothesis is considered speculative without multiple rounds of testing.

It is unclear how Watson envisions hypothesis generation in science. All science begins with an observation, and an attempt to account for it, generally using a research paradigm or model, then testing that account. In this regard, evolutionary psychology is standard science. Watson seems to believe we can know what is true or false before we've even started. See also point 19.

An evolutionary hypothesis constrained by the conditions of the past is often easy to test and falsify because that feature leads to some very specific predictions. If people could learn to write as easy as they learn to speak, the notion that language is an adaptation would be in serious trouble because writing is a recent invention. The hypothesis of language as an ancient adaptation is easily falsifiable.

`16:00` Watson recounted how neuroscientist V.S. Ramachandran published a "satire" study "Why gentlemen prefer blondes."

34

Ramachandran published his paper in a non-peer reviewed, non-evolutionary psychology journal *Medical Hypotheses*. If his "satire" submission was so indistinguishable from "real" evolutionary psychology, why not publish in a mainstream peer-reviewed evolutionary psychology journal?

Medical Hypotheses defines itself as a space for unfounded and out-there ideas. The journal's own website reveals: "The journal will consider radical, speculative and non-mainstream scientific ideas provided they are coherently expressed." About the journal Don Symons said, "If Medical Hypotheses is peer reviewed, it must be by chipmunks" (Symons, 2000).

`16:50` Watson portrayed V.S. Ramachandran as a rebellious iconoclast, calling him the "honey badger of science," and saying, "he don't give a fuck." One might imagine that Ramachandran would find the publisher (*Medical Hypotheses*) of his "satire," the purported "dupee" worthy of derision, having foolishly printed his joke paper. The journal is indeed questionable, having raised eyebrows when it published a paper in 2009 by AIDS denialist Peter Duesberg, denying that HIV causes AIDS (Enserink, 2010b).

35

Watson failed to tell her audience that Ramachandran has published a total of fifteen times in *Medical Hypotheses* and sits on their editorial advisory board. In spite of everything, Ramachandran praised the journal as a "unique and excellent venue for airing new and valuable ideas."

Ramachandran's scientific skepticism is perhaps unfitting of the fawning praise Watson gave him, to the detriment of her audience (Enserink, 2010b).

17:25 Watson cited Satoshi Kanazawa's work as an example of evolutionary psychology.

36

Satoshi Kanazawa is a disgraced outlier, roundly criticized from within evolutionary psychology (BBC news, 2011). To wit:

Watson cited an interview Kanazawa did for *The Sun*—a tabloid newspaper. In it, Kanazawa cited no original peer-reviewed research.

In 2011, he was fired from his job blogging at *Psychology Today*.

68 evolutionary psychologists issued a statement condemning his work on the basis of its poor quality and dishonest methods. It is titled "Kanazawa's bad science does not represent evolutionary psychology" (Alvergene et al., 2011).

According to the above statement, 24 critiques involving 59 different scientists have been published in peer-reviewed journals of Kanazawa's work. Kanazawa has not published a full length reply in an academic journal to the many critiques since 2002, showing his disengagement.

35 leading minds in psychology, including evolutionary psychology, wrote a total deconstruction of his research model and published it in commentary in *American Psychologist* (Penke et al., 2011).

His own employer, the London School of Economics, forbade him (as punishment) from publishing in non-peer reviewed outlets for a full year and distanced themselves from some of his work (BBC News, 2011).

20:07 During the lead-in to the interlude about Cindy Meston and David Buss's *Why Women have Sex* Watson said *[W]omen hate sex. Science has proven it ... women hate sex unless they're using it to get money or babies.*

37

During the several minutes Watson spent on the topic, no researcher or paper is ever cited that claims this, and several claim the contrary. On Amazon's copy about the book, the promo explicitly includes "pleasure" as one reason women like sex. Buss and Meston's research clearly shows pleasure is important to both sexes. Watson noted that their research shows that both men and women cite pleasure as a reason at minute 21:45. It is baffling why Watson would begin this section *[W]omen hate sex. Science has proven it,* by citing two researchers whose work demonstrates many stereotypes about men and women are false (Castleman, 2010).

In discussion of the top 10 reasons men/women have sex, the top for both was that they're attracted to the person, and there was not much difference between genders according to Cindy Meston in a presentation she and Buss gave (Buss and Meston, 2009).

21:00 Regarding Meston and Buss's book, Watson remarked, *They bravely went and interviewed a thousand white middle class women to figure it out...*

38

The women polled were diverse in ethnicity, socio-economic status, and came from the US, Canada, New Zealand, Israel, China, and Australia. Again, it is unclear why Watson wished to cast doubt on mainstream research that supports her own positions on gender and sex stereotypes (Meston and Buss, 2007).

21:59 *Apparently only women are the mysterious creatures that need an entire book to figure out why they like sex...*

39

Watson changed the language to a more disparaging characterization "why women *like* sex" from why women *have* sex, in spite of it being the title of the book. Those are different questions, and no one wonders why women (or men) like sex.

Most psychology research has a relatively narrow focus. Researchers chose this topic because it interested them, much the same as any researchers in any field. Trying to understand one particular gender identity, age, race, class, or any other sub-group better is a perfectly legitimate goal unless there is evidence of bias, but no such evidence was presented. Social scientists have often been (rightly) criticized for unduly focusing on men as normative. Surely, we should applaud Buss and Meston for giving much needed focus to female sexuality and doing justice to its complexity and nuance.

22:45 *There have been other studies where women didn't really seem to figure into it at all. [Gendered differences in receptivity to sexual offers] is a particularly fun and horrific one.*

40

It is unclear in what sense women did not "figure into" this study, or what makes it "horrific." Equal numbers of the experiment's subjects and research confederates were male and female. The authors did not favor one gender or the other in the design, discussion, or findings. The paper was co-authored by a woman, Elaine Hatfield.

22:52 *There are a number of studies based on this idea men appear to enjoy casual sex way more than women do. And women, of course, again, tend to only want to have sex when they get a husband out of it, or babies or money. So they take this as a given and they do studies like this [indicates Clark and Hatfield 1989 on the slide] in which they set out to prove it as a fact then make up a story about how our Pleistocene era brains are somehow responsible for this.*

41

The idea that Clark and Hatfield "set out" to establish some evolutionary account could hardly be more mistaken. The purpose of the study was to try to arbitrate between competing theories. They recounted dominant theories of the time, both the evolutionary and the cultural account without praising or disparaging either:

According to cultural stereotypes, men are eager for sexual intercourse; it is women who set limits on such activity.... No experimental support for this hypothesis exists, however. In this research, we will report an experimental test of this proposition (Clark and Hatfield, 1989).

42

Clark and Hatfield were social psychologists, not evolutionary psychologists. They showed due consideration of sociological explanations. They wrote in the conclusion: *Of course, the sociological interpretation—that women are interested in love while men are interested in sex—is not the only possible interpretation of these data. It may be, of course, that both men and women were equally interested in sex, but that men associated fewer risks with accepting a sexual invitation that did women. ... [A]lso the remnants of a double standard may make women afraid to accept the man's invitation* (Clark and Hatfield, 1989).

`23:20` *In this particular study, a group of attractive young white people were sent out into the public to invite members of the opposite to have sex with them.*

43

The paper says nothing about the races of the confederates. The research confederates ranged from "slightly unattractive to moderately attractive." Attractiveness was not a requirement, and the authors concluded that it also had no effect on the results in any case (Clark and Hatfield, 1989).

`23:20` *...a group ...were sent out into public to invite members of the opposite sex to go to bed with them. On the spot. Like, let's go back to my apartment right now and have sex.*

44

Participants were asked one of three questions, none of which was "will you have sex with me right now?" The questions were: "Would you go out with me tonight?", "Would you come over to my apartment tonight?", and "Would you go to bed with me tonight?" The study took place in the daytime between classes so "tonight" meant later, not now (Clark and Hatfield, 1989).

`23:32` *Across two studies, 69% and 75% of men accepted the offer and no women did. So obviously, women hate sex.*

45

In the discussion of expected results, it is clear the researchers did not know exactly what to expect. They outlined three possibilities; one in which both sexes were more receptive than expected, another in which one or the other sex was more so, and one matching stereotypical views. From the paper: *It may be, however, that men and women are not so different as social stereotypes suggest. Again and again, researchers have found that while men and women expect the sexes to respond in very different ways, when real men and real women find themselves caught up in naturalistic settings, they respond in much the same way* (Clark and Hatfield, 1989).

46
Nowhere in the paper did the authors say or imply that women hate sex.

25:33 Watson suggested an alternate explanation for the conclusion of the sexual offers studies and implied the authors had not considered it: the threat of rape.

47
The 1989 Clark and Hatfield paper stated as part of its conclusion that the risk of assault is greater for women, and that this could help explain the findings (Clark and Hatfield, 1989).

48
In a presentation condemning the concoction of untestable "just so stories" Watson asked her audience to disregard the findings of a dozen scientific studies across multiple countries and decades in lieu of her ad hoc story about female psychology for which no evidence is given other than a single assault case from the news.

49
Her explanation was contradicted by two different papers she cited herself. Guéguen 2011 (see 24:30) found that when

propositioned by an attractive male, 57% of women agree to go to their apartment—just the activity Watson said women were too fearful to do (Guéguen, 2011). At point 58 Watson favorably referred to Conley 2011 which also contradicted this point: *...perceived danger variables did not predict acceptance of the [sexual offer] for women or for men.*

27:16 *[O]ne study looked at speed dating. What they found is when they let the women make all of the decisions about who they got to talk to ... they were much more likely to be cocky and non-selective as the men ... they were slutty.* Watson did not provide any details identifying the study. It appears to be "Arbitrary Social Norms Influence Sex Differences in Romantic Selectivity" (Finkel and Eastwick, 2009).

50

Women did not make any decisions about who they got to talk to. That is not how speed-dating works. Every person of one sex talked to every person of the opposite sex (only heterosexual events were considered). This is why the authors spoke of who was "rotating" and not "selecting."

51

By "much more likely" to be "slutty," Watson meant that the women said yes to a date 45% of the time (when approaching) compared to 43% of the time (when men were the ones approaching) (Finkel and Eastwick, 2009).

52

The authors' conclusion is contradicted by their own data. When men were not "rotators" they said yes 50% of the time, compared to women who said yes 45% of the time when not "rotators," a significant sex difference consistent with the evolutionary perspective.

53

This speed dating study has as research subjects young, likely mostly white, middle class college undergraduates. Watson vocally criticized this practice several times in her talk (see 38, 43, 70).

54

Harvard psychologist Steven Pinker noted that the artificial conditions of the experiment limited the meaningfulness of the results.

> They never acknowledged a crucial point: that they are studying the special situation in which women have already consented (or have been instructed—they give no details about this part of their methods) to approach strangers for possible romantic or sexual opportunities. That is, they ignored the stage in the mating process in which the largest sex differences are predicted to occur, namely whether people are willing to approach strangers for possible romantic opportunities in the first place (Pinker, 2012).

`27:40` *[I]n a survey of two thousand single women ... researchers found 82% agreed to at least one casual sexual encounter.* Watson gave no source or means of identifying the survey, but the details match a 2011 article for *Self* magazine, as mentioned, a clone of *Cosmopolitan* (Self Magazine, 2011).

55

Previously Watson criticized *The Telegraph* for publishing a study conducted by an actual researcher because it was paid for by a business, saying, *All of the best studies I find are commissioned by shopping centers. This is actually marketing disguised as science* (see point 13). Here, 23 minutes later, she used a survey created by a magazine expressly to market to its readers as legitimate evidence supporting a point she is making—and she intended it as rebuttal of peer-reviewed scientific research.

56

Self magazine is clearly marketed to young, white Americans. The survey surely skews heavily that way, making it just the sort of demographic sample Watson criticized at points 38, 43, and 70.

57

The evolutionary psychology theory of differing sexual selectiveness among the sexes does not predict women would not engage in casual sex, but that they would do so under different circumstances than men some statistically significant part of the time. Men, conversely, could not have evolved any preferences for casual sex if there was no one to have casual sex with.

`27:52` Citing Conley 2011, Watson said *[R]esearchers asked subjects how they viewed the people who were approaching others for sex in that 1989 study. Women and men both agreed that the women proposers in the experiment were intelligent, successful, sexually skilled much more so than the men who were asking for sex. And the researchers are thinking that that's because we see women as being more passive and shy and so we assume that a woman who is confidently asking for sex is really good at sex.*

58

The authors admitted their finding is far less valid than Clark and Hatfield 1989 because that study was naturalistic, observing actual human behavior. Conley's method was a written survey asking subjects what they would do or what

234

their opinion is, not observing what they actually do. The subjects know that they are being studied and are biased accordingly (Conley, 2011).

59

Subjects affirmatively answering the questions about sexual skill is not strong evidence that sexual skill is part of anyone's decisions about casual sex and is disconfirmed by much evidence (e.g. the lack of discussion with regard to sexual skill indicators in personal ads or dating scenarios). It is also possible that people elevate the expected sexual capabilities of people they choose to have sex with for other reasons.

60

In the only part of Conley 2011 approaching naturalistic, subjects were asked about real casual sexual encounters of theirs. Here, Conley reported an enormous sex difference: 73% of men versus 40% of women reported accepting offers.

61

Conley disagreed with a prevailing evolutionary hypothesis, "sexual selection theory," but favored a different evolutionary hypothesis which also rests on gender differences Watson has attacked. Conley wrote

> [T]he current findings support a theory with evolutionary foundations (i.e., pleasure theory) in showing the primacy of pleasure in sexual decision making.

Conley cited Armstrong et al. 2012's research on pleasure theory, which includes such findings as

> [Women] who were interested in a relationship were about a third more likely to orgasm and more than twice as likely to

say they enjoyed the hookup. (Armstrong, England, and Fogarty, 2012)

In other words, Watson used her own citation to argue against evolutionary psychology when that study states plainly that it hypothesizes a theory with "evolutionary foundations." The theoretical basis of the paper is full of assumptions of gender differences that Watson disclaims as sexist and nonsensical. Whether such differences exist or not, it is unethical to misrepresent the research which, in turn, mislead the audience.

29:26 Watson said that the study (she does not provide any citation, but it could only be (G. Miller, Tybur, and Jordan, 2007)) suggesting strippers make more when they're ovulating has serious methodological problems. She remarked that there are only 18 participants, data was collected over 60 days, that the strippers only reported data on 27% of days that they were supposed to, that all of them worked in the same club and this would therefore bias the data if there were particularly busy days (such as a convention being in town). Watson also suggested that it is obvious why earnings would decline during menstruation citing the physically unpleasant aspects.

62

The small sample size limits the generalizability of the study, but small sample sizes are sometimes unavoidable. The fact that the participants did not report data as often as the researchers wanted is only pertinent if the authors did not collect adequate data for each phase of the cycle, or if there was some patterned bias in reporting. Neither possibility could explain the cycle effects observed. The co-location of the participants could only be relevant if the women were cycling together and spikes in business only occurred during the fertile period. One might indeed assume the week of menstruation could affect job performance (the authors said they expected it), but that was not the most important aspect of the data. The study was about ovulation (egg release/fertility) not about menstruation. Strippers made more money during peak fertility than they did any other time, including the non-menstrual phase. Additionally, participants using hormonal birth control made substantially less money during their fertility phase. None of Watson's objections mentioned or refuted these findings.

Lastly, Miller et al. built on solid existing research, citing thirteen previous studies documenting apparent detectability of ovulatory cues (Gangestad et al. 2002; Grammer et al. 2004; Haselton and Gangestad, 2006; Haselton et al. 2007; Havlicek et al. 2006; Kirchengast and Gartner, 2002; Krug et al. 1999; Kuukasjarvi, 2004; Manning et al. 1996; Pillsworth and Haselton, 2006; Provost et al. 2008; Roberts et al., 2004; Singh and Bronstad, 2001; Symonds et al. 2004). The basic finding has been supported by subsequent research (Alvergne and Lummaa, 2010; Fink et al. 2012; Haselton and Gildersleeve, 2011).

`31:01` Watson cited a 2007 study testing for a link between women's gait and their ovulatory status (Watson provided no citation, but it could only be Provost et al. 2008) Watson said, *the results were no. And so their conclusions was that that must mean women are trying to disguise their fertility to deter unsuitable partners.*

63
Provost et al. *did* find an effect in one of their two experiments. Men's gait preferences *did* change with ovulatory cycle status, just not in the expected direction

(men showed a preference for gait of women in non-fertile phase).

64

Watson misrepresented the authors when she said they *concluded* what this *must* mean. She was referring to speculative discussion of what the results might mean in the section of the paper titled "Discussion" not "Results." They wrote,

> This finding contradicts the face research... it is possible that faces and gait present different information because of the intimacy with which the stimulus is viewed. For example, faces can only be seen in a fairly close encounter, whereas gait patterns can be seen from a large distance.

The authors reported the effect, but made no claim that the cause is proven. It is standard practice in any good research publication to attempt to interpret unexpected results. In fact, it is often very important to do so because it can help guide further research to either rule out possibilities or to design improved future studies. Furthermore, Provost et al's speculation that close-contact signals (facial cues) should differ from wider-rage cues (walking gait) is perfectly possible, whether true or not.

A subsequent experiment, Fink et al. 2012 did find an effect of fertility phase on perceived attractiveness of gait in a more controlled study building on Provost. Fink et al. found a strong effect of ovulatory phase. Nonetheless, they wrote in the conclusion:

[I]t is also clear that particularly with regards to the role of body movement in this context, further research is needed before we can confirm or reject the perspective on women's

body movements as being another physiological by-product of the ovulatory cycle and whether or not men are obtaining information regarding fertility from them.

That's because these studies are initial forays into a new area. Speculation, guesses, hypothesis refinement, and generally lots of experiments with negative or uncertain results are normal in science. Everyone involved in the research understands the findings are tentative and amount to little yet. It is again unclear what it is that Watson takes science to be.

65

This is another paper Watson cited which did not find support for a particular evolutionary hypothesis in spite of her insistence that such hypotheses cannot be tested.

31:24 *In 2010 researchers* [Watson does not identify which, but from the details she gave it could only be Miller and Maner, 2010] *found that in a group of 38 men, the men found that a woman they interacted with was least attractive when unbeknownst to them she was ovulating ... so that was explained as the men fooling themselves into thinking they weren't attracted to her so that they could maintain relationships with their girlfriends. This was, according to the researchers, because evolution would favor a man who could stay with a woman long enough to bear children. Which goes against all the other evolutionary psychologists who just think that men are there to spread their magical seed.*

66

Watson failed to mention that the study included both men who identified as single and who identified as "in a committed relationship." Each group had divergent opinions of the woman (the single men rating her more attractive, the non-single men, less attractive) but *only* during ovulation, which is quite striking (Miller and Maner, 2010).

67

The notion that this study "goes against" a prevalent evolutionary psychology staple of male promiscuity is either naïve or a deliberate distortion, since as no evolutionary psychologists defend this cartoonish view. To observe that males are comparatively more promiscuous and less discriminating about sexual partners is not to claim that they are incapable of, or do not adaptively benefit from, long term relationships. Humans, male and female, have a diverse array of mating strategies which may be more or less adaptive based on an individual's characteristics, resources, options, and environment. Among those is pair-bonding, which evolutionary psychologists identify as providing unique and important adaptive benefits for males. See Gonzaga et al. 2008 for discussion.

68

The Miller and Maner paper does not contradict findings on male promiscuity for other reasons. Some useful context is required: the woman was not permitted any kind of signal of flirting or sexual availability. She wore no make-up, her hair always pony-tailed, her clothing was always a t-shirt and jeans, and flirting behaviors discouraged and monitored-for. The woman was selected based on her "average" attractiveness and the interaction with each male was a single 20-minute session of "cooperative tasks." In other words, the situation was very nearly as unsexy and uninviting as it could possibly be. It is entirely consistent with evolutionary psychology theory that context is important and in this situation, when deliberately removed from any overt sexual cues we should not find it at all strange that committed men were not lured astray. We should also find it all the more astonishing that invisible cues to fertility nonetheless made them down-grade their ratings, while the single men upgraded them (Miller and Maner, 2010).

In this and in the preceding points, Watson has tried to create the impression that disagreements among researchers proved that the field is sexist or dishonorable. In any healthy science, particularly in new research areas such as those quoted above, there are rightly robust disagreements at times.

`32:10` *In 2009 researchers wanted to know if women were more racist when they were ovulating. ... [B]ecause they're running out of shit to study while women are ovulating.*

69
The punch line of Watson's joke (which did get laughs) is that the 2009 study about racism is frivolous. Her attitude makes it appear unlikely she knew that Navarrete et al. chose to focus on ovulation because women have been uniquely vulnerable to sexual coercion and assault throughout history, and the effect studied (if real) may represent a coping mechanism. Her flippant dismissal of reputable peer-reviewed research endeavoring to produce new knowledge about the social tragedies of racism and sexual coercion would be appalling even if she had not ridiculed the field as racist and sexist five minutes later (see point 85).

Her remarks might have made sense if she knew this literature fairly well, and knew it to be empty of validity, but she proved this was not the case as she misconstrued the details and admitted to not knowing about two important papers on the topic (See points 70, 71).

32:23 *[A]nd what they found was that women ... might be implicitly more racist. And that study involved 77 white psychology students at university and has never been replicated.*

70

Black and white participants were recruited, but only ten black women met the qualifying criteria (non-pregnancy status, not using oral contraceptives, answering all survey items, et cetera) and had to be omitted from the analysis because n = 10 would be too small to have much statistical power. As a first study of this kind, the scope is limited. There is also nothing whatsoever wrong with one study focusing on the racism of white people (or any other one ethnicity) as such findings can be valid and important (Navarrete et al, 2009).

71

Despite Watson's very emphatic "...and has never been replicated," the finding was replicated in four experiments documented in two subsequent papers (McDonald et al., 2011; Navarrete et al., 2010).

34:24 Watson introduced a *NewScientist* article on color perception by Anya Hurlburt and Yazhu Ling by calling it another example of "evolutionary psychologists trying to support a shitty stereotype about women," namely the preference for the color pink. (Hurlbert and Ling, 2007; Khamsi, 2007)

72

Neither author is an evolutionary psychologist, or even a social scientist for that matter. Neither has any academic background in evolution-based behavioral science nor do they publish in such journals.

242

73

Neither author is a man. In a talk about lax regard for evidence and for bias by evolutionary psychologists and/or journalists covering them, Watson is asking her audience to believe these accomplished women are determined to oppress their own gender—because Watson did not like their speculative remarks. Dr. Anya Hurlburt, for example, is director of the Institute of Neuroscience at Newcastle University and holds a degree in physics, a Master's in physiology from Cambridge, a PhD in Brain and Cognitive Sciences from MIT, and an MD from Harvard. Other than disagreeing with one of her dozens of published papers, Watson offers no evidence to support her deeply insulting claim that Dr. Hurlburt is abusing science in service of self-targeting sexism.

`34:27` *Here is a study that shows that pink is a girl's color because women were Pleistocene gatherers who had to be able to tell when berries were ripe.*

74

The study does not show what Watson asserts, nor was it meant to. It was meant to produce objective data about color preferences between cultures and genders, which it did, and those findings are consonant with field literature (Hurlbert and Ling, 2007).

`34:45` *We have substantial evidence that up until the 40s or so pink was a boy's color and blue was a girl's color.*

75

Cecil Adams of *The Straight Dope* has pointed out that prior to the first half of the 20th century, there simply was no dominant cultural idea of a color "for" boys or girls. Young boys and girls dressed in white or a variety of colors with no apparent sexual connotation whatsoever. The matter seems to have been settled by decades-long debate, but there may have been no consensus at all, ever, that pink was the color for boys (Adams, 2008).

76

Watson was confused about what an evolutionary psychology explanation means. To say that an observed psychological feature has some basis in our evolutionary past is not to say that it has had the same manifestation at all times and in all circumstances. People consume pornography partly because of the evolutionary link between sex and adaptive fitness, not because pornography has always existed and always been available to humans (the opposite would be true, actually). Watson and the researchers she ridiculed could actually both be correct at the same time.

77

The authors reported that both sexes and both English and Chinese ethnicities overwhelmingly prefer blues to any other color. This is not an impressive effort at supporting any stereotypes (Hurlbert and Ling, 2007).

35:15 *The study disproves itself because part of it was done in China, and what they found in China is that men also prefer pink.* Everything in this remark is incorrect.

78

No part of the study was done in China. It was conducted in the UK and included ethnically Han Chinese participants who had recently relocated to the UK.

79

The study did not measure male (or female) preference for pink, but rather the shape of a preference curve across the blue-yellow and red-green hue contrasts. The sex difference in which the red-green curve "leans" positive for women and negative for men was found for both cultures. The authors wrote,

> Only the "red–green" weights show a consistent sex difference across all populations. On average, all males give large negative weight to the [red-green] axis, whereas all females weight it slightly positively (sex difference $p < 0.00001$). That is, females prefer colors with "reddish" contrast against the background, whereas males prefer the opposite (Hurlbert and Ling, 2007).

80

The authors *did* expect to see some cultural influence, including the Chinese belief that red is lucky. Watson implied they were surprised and hamstrung by the finding. They weren't. From the *NewScientist* article Watson displayed on screen: *The idea of testing the two groups was to separate out whether culture or biology might influence gender preferences for color* (Khamsi, 2007). It is also worth noting that Watson implied that she has a more sensible opinion on color preference and culture than Yazhu Ling, a highly educated Chinese-born woman who has lived and worked in China, the UK, and the US and has made human color perception her career.

81

As in point 64, Watson has confused a paper's conclusion (the thesis supported or unsupported by the data) with speculation about the meaning of that conclusion. Researchers are expected to speculate about findings in published papers. In this particular paper, a reader may note such a tone by the phrase "We speculate that...," which the authors wrote on page 624 (Hurlbert and Ling, 2007). Watson said the paper "proves" and "shows" that which was merely speculated (or entirely absent).

35:40 Watson suggested that the stereotype that men cry less was invented and perpetuated by evolutionary psychologists.

82

The evidence is that a reduced male tendency to cry (or to appear to cry) is not cultural. A 2011 study of 37 nations including ones in Europe, North and South America, Asia, Africa and Oceania with a total of 5,715 subjects found that in all places women report a stronger tendency to cry and to have done so more recently. Analysis of the 74 comparisons (2 measures x 37 nations) showed the same effect size for sex in all but three. This is very substantial evidence sex plays a significant role independent of culture (van Hemert, van de Vijver, and Vingerhoets, 2011). Lombardo et al. conducted a study on gender and crying in 1981, then again in 1996 using diverse subjects. Although many researchers have noted that the culture of the US has changed udring that time, the two studies reported the same results (Lombardo, Cretser, and Roesch, 2002).

36:08 Watson said that the idea that a woman's place is the home is novel to the industrial era and that evolutionary psychologists are bent on looking for reasons to support the stereotypical view.

83

In point 30 Watson cited Kuhn and Stiner (2006) to support her contention that present small-scale societies do not necessarily resemble past ones. However, Kuhn and Stiner were arguing that that point in the distant past was the start of division of labor for *Homo sapiens,* around 40,000 years before the industrial revolution. Watson may wish to reconsider citing papers that contradict her position, or at least not to ignore primary points in a paper she alleged that evolutionary psychologists had ignored.

84

Watson provided no evidence for the outlandish assertion of active sexism. Here is what prominent evolutionary psychologists actually say:

> Nothing in evolutionary theory privileges males over females, however, nor does evolutionary theory prescribe social "roles" for either sex. Are ovaries superior to testicles? The question is meaningless. Are male mate preferences superior to female mate preferences? The question is equally meaningless. -Edward Hagen (2004)

> [E]volutionary explanations of the traditional division of labor by sex do not imply that it is unchangeable, "natural" in the sense of good, or something that should be forced on individual women or men who don't want it. -Steven Pinker (1997)

> We share the view that men's historical control of power and resources, a core component of patriarchy, can be damaging to women in domains ranging from being forced to endure a bad marriage to suffering crimes such as genital mutilation and "honor killings" for perceived sexual infractions. -David Buss and David Schmitt (2011)

36:58 Watson intimated that evolutionary psychologists are *using bad science to keep women and minorities down [which is]* is *nothing new*, and that *not much has changed in a hundred years.*

85

There is no evidence that the evolutionary psychology research program is based on subjugation of anyone, and much to the contrary. Refutations in point 69 are similar. The claim of oppressing minorities is especially bizarre. A fundamental claim of evolutionary psychology is that the mind's structure is basically pan-human, an idea which would undermine any biological justification of prejudice on the basis of class or race. See Buss and Schmitt (2011) for discussion of evolutionary psychology and feminism.

`38:19` Watson said that evolutionary psychologists assert that "men evolved to rape" in order to justify rape with "it's natural for men to rape" and/or that people will construe it this way.

This part of Watson's talk was vague and unclear. To recap a bit, in minute 36 Watson explained "traditional" sex roles are an artifact of the industrial revolution: *... and then when the industrial revolution came around men started working in the factories leaving women at home to take care of everything else. So now evolutionary psychologists ignore all that and pretend that women's place is in the home and they look for reasons to "scientifically"* [makes air quotes] *support that. So it's all good to chuckle at bad science but what is the harm in bad science that perpetuates stereotypes...*

At that point, the topic was "evolutionary psychologists" and their "bad science" that supports stereotypes. Next Watson said that this is "nothing new" and discussed how biology textbooks were used (by whom?) against early twentieth century women fighting for suffrage, adding "and

248

not much is changed in a hundred years." Watson then quoted a FOX news contributor, Reverend Jesse Lee Peterson who made a misogynistic statement about "evil reign[ing]" as a result of women "taking over." At that moment, it was not clear what point Watson was trying to make. Peterson is a right wing religious ideologue citing no textbook or science. Perhaps she cited Peterson to speak to the harm of stereotypes. On the following slide, Watson continued:

So, there are a lot of ways stereotypes screw us up. One is that they are sometimes used as the basis to limit our own rights. It's also used to excuse predatory behavior like "men evolved to rape," maybe you've heard of that one. Newsflash, it's bullshit, but it was used as a sort of... well you know it's natural for men to rape therefore we don't really need to look into the ways that we can change our culture to stop men from raping, it's natural.

Who used it? Who is making this claim? To clarify, consider Watson's answers during an interview after giving the same talk at the World Skeptic's Congress in Berlin in May of 2012:

Watson: *Yeah, basically I used my talk as an opportunity to slam evolutionary psychology for half an hour, cause–*

Interviewer: *Good [unintelligible]*

Watson: *Yeah, well, it's been building up for a while. I just get so tired of seeing. "Women evolved to shop," "Women evolved to like the color pink," "Women evolved to be terrible at math and logic."*

Interviewer: *Oh yeah, and "Men are evolved to rape."*

Watson: Yeah. "Men evolved to rape." Uh, yeah. I mean, the thing is, once you look past the headlines and actually look at the studies, what you see over and over and over again is pseudoscience being passed off as science. You know, they have tons of assumptions that they don't support with the evidence, and they make up just-so stories that seem to fit the facts. And it only ends up reinforcing stereotypes, which does harm to all of us.

Five minutes later in the interview, she stated,

I think there are people who hold misogynist, racist, bigoted ideas, but they value science, and so they will seek out what they consider science in order to support their prejudice. And it's been happening since the beginning of time. I mentioned during my Q&A that evolutionary psychology is not a new thing. It's becoming more and more popular in the last few years, but it's actually evolved from other things, like Social Darwinism, which, you know, got into a lot of trouble over eugenics and things like that.

Now it is clear that Watson could only have been referring to evolutionary psychology researchers and those who support them. She explicitly said "when you look past the headlines" and identified the field as Social Darwinism reborn. When she spoke of those who believed "men evolved to rape" is some sort of justification, it must be concluded that she means evolutionary psychologists and the lay people that support or believe them.

86
Watson provided no evidence for the claim that the evolutionary psychologists studying rape (or proponents of evolutionary psychology) attempted to morally justify it in

any way, shape or form. No research or researcher was quoted or cited.

87

The evolutionary psychology of rape informs that rape is a *more* heinous violent crime than other types of assault. Commenting on Thornhill and Palmer's book on the subject, Tooby and Cosmides wrote,

> "Thornhill and Palmer argue that women evolved to deeply value their control over their own sexuality, the terms of their relationships, and the choice of which men are to be fathers of their children. Therefore, they argue, part of the agony that rape victims suffer is because their control over their own sexual choices and relationships was wrested from them." (Tooby and Cosmides, 2000)

88

There is no monolithic view on the truth of the hypothesis that rape could be an adaptation. David Buss and David Schmitt wrote in a 2011 paper *Evolutionary Psychology and Feminism*, "We concur with Symons's 1979 summary that the then-available evidence was not 'even close to sufficient to warrant the conclusion that rape itself is a facultative adaptation in the human male' (Symons 1979, p. 284). We believe that his conclusion is as apt today as it was then." These are very influential people in evolutionary psychology rejecting the claim which Watson said evolutionary psychology exists to promote (David Michael Buss and Schmitt, 2011).

89

42:16 Watson stated that stereotypes reduce minority interest in such things as skeptical events and organizations. She described a Stanford study which demonstrates the harmful effect of stereotype threats. Stereotype threat may be a worthy area of study, but it is not especially relevant to Watson's topic, nor is any connection between the two created beyond mere assertions and confused or mistaken references.

47:30 In response to a person in the audience asking, "Is there any good evolutionary psychology?", Watson replied:

Prooooooobably? I'm guessing yes, but it's so boring, because you can only make it interesting if you make up everything. Because, really, good evolutionary psychology would be more like, "Well, we don't really know what happened in the Pleistocene, and we have no evidence for this, but maybe this. It's not the sort of thing that makes headlines. So if there is good evolutionary psychology, it's not in the media, and therefore, it might as well not exist as far as the general public is concerned.

90

Watson appeared unaware of the uncontroversial fact that much of evolutionary psychology is mainstream, reputable and well-tested. She admits to not being familiar with anything that is not hyped in the media (in spite of insisting in a previous interview that it is pseudoscience "once you look past the headlines") and it turns out she is also not familiar with evolutionary psychology that *is* in the media—including studies she has cited herself. Watson has gotten the facts exactly wrong. It's very easy for studies critical of evolutionary psychology to get attention in the media.

There are two ways to take Watson's certainty that there is no "good evolutionary psychology in the media," and both are empirically false. The first is in terms of studies that literally fit her stated criteria, that is, those that question or seek to disprove evolutionary accounts as overreaching or mistaken. Watson herself favorably cited research investigating evol utionary psychology hypotheses. Assuming Watson does not believe that they "made everything up" to be interesting, we may conclude that she must think of them as good studies. She may not think of them as evolutionary psychology studies, but since they specifically and deliberately investigate and comment on the evolutionary account, they categorically are. Do they have trouble getting media coverage as Watson said?

At minute 27:16 Watson mentioned a study by Eli Finkel and Paul Estwick (see points 50-54). It was featured, favorably, in both *Nature* and *The New York Times* (Arnquist, 2009; Kaplan, 2009). Journalist and writer Dan Slater wrote a *New York Times* op-ed about the work of Finkel and others. He called his anti-evolutionary psychology polemic "Darwin was Wrong about Dating" (Slater, 2013). Watson cited work by Steven Kuhn at minute 14:47, again favorably. Kuhn's research got good press from *National Geographic News*, and ScienceDaily ("Gendered Division Of Labor Gave Modern Humans Advantage Over Neanderthals," 2006; Lovgren, 2006).

Watson told her audience to read Cordelia Fine. Fine's latest book was well reviewed by the *Guardian* and the *New York Times* (Apter, 2010; Bouton, 2010). Biologist and behavioral ecologist Marlene Zuk wrote the book *Paleofantasy*, specifically to repudiate pseudoscientific myths about our evolutionary past. Zuk says almost verbatim the sort of thing Watson described, in this case about the human diet. Zuk explained that there is too much variation in time and space, "We simply ate too many different foods in the past and have adapted to too many new ones, to draw such a conclusion" (Zuk, 2013). Zuk and her work were featured by *Slate, Salon, The Guardian*, and *NewScientist* (Forbes, 2013;

George, 2013a, 2013b; L. Miller, 2013). It seems that "good" studies per Watson's own description have no trouble at all getting media coverage.

The second and less myopic question relevant to Watson's curious statement is, does good evolutionary psychology in general make it into the media? Absolutely. Watson mentions a few books, so let's start with a short list of reputable, well-reviewed pop evolutionary psychology books.

The Moral Animal: Why We Are, the Way We Are: The New Science of Evolutionary Psychology by Robert Wright
The Selfish Gene by Richard Dawkins
Evolutionary Psychology: The New Science of the Mind by David Buss
How the Mind Works by Steven Pinker
Religion Explained: The Evolutionary Origins of Religious Thought by Pascal Boyer
The Adapted Mind: Evolutionary Psychology and the Generation of Culture by Jerome Barkow, Leda Cosmides, and John Tooby
In Gods We Trust: The Evolutionary Landscape of Religion by Scott Atran
The Red Queen: Sex and the Evolution of Human Nature by Matt Ridley

Good reporting can be found many places, such as the *Scientific American* and the *Smithsonian* magazines (Ward, 2012; Wayman, 2012; Weierstall, Schauer, and Elbert, 2013). My own work was reasonably well covered by *National Geographic* and the *Smithsonian* (Nuwer, 2013; Zimmer, 2013). Great evolutionary psychology discussion is regularly found at *Edge.org*. Many evolutionary psychologists publish blogs including Diana Fleischman and Robert Kurzban who writes at *Mind Design* for *Psychology Today* as well as the blog for the journal *Evolutionary Psychology* (Fleischman, 2013; Kurzban, 2013a, 2013b). In fact, there are eighteen blogs at the *Psychology Today* website with a focus on evolutionary psychology. The evolution magazine *This View of Life,* edited by David Sloan Wilson often has great articles (Wilson, 2013).

Good science reporting for nearly any field is frustratingly rare, but evolutionary psychology is no more or less so than any other.

References

Adams, C. (2008). The Straight Dope: Was pink originally the color for boys and blue for girls? *StraightDope.com.* Retrieved October 12, 2013, from http://www.straightdope.com/columns/read/2831/was-pink-originally-the-color-for-boys-and-blue-for-girls

Alcock, J. (2000). Misbehavior How Stephen Jay Gould is wrong about evolution. *Boston Review.* Retrieved May 28, 2013, from http://www.bostonreview.net/BR25.2/alcock.html

Alvergene et al. (2011). Kanazawa's bad science does not represent evolutionary psychology. *Evolutionary Psychology.* Retrieved May 28, 2013, from http://www.epjournal.net/wp-content/uploads/kanazawa-statement.pdf

Alvergne, A., and Lummaa, V. (2010). Does the contraceptive pill alter mate choice in humans? *Trends in Ecology and Evolution, 25*(3), 171–179. Retrieved from http://www.sciencedirect.com/science/article/pii/S016953470900263838

Apter, T. (2010). Delusions of Gender: The Real Science Behind Sex Differences by Cordelia Fine. *TheGuardian website.* Retrieved May 28, 2013, from http://www.guardian.co.uk/books/2010/oct/11/delusions-gender-sex-cordelia-fine

Armstrong, E. A., England, P., and Fogarty, A. C. K. (2012). Accounting for Women's Orgasm and Sexual Enjoyment in College Hookups and Relationships. *American Sociological Review, 77*(3), 435–462. doi:10.1177/0003122412445802

Arnquist, S. (2009). Testing Evolution's Role in Finding a Mate—NYTimes.com. *The New York Times website.* Retrieved May 28, 2013, from http://www.nytimes.com/2009/07/07/health/07dating.html?_r=0

Barnett, L. (2009). Primitive lot, these shoppers. *Express.co.uk.* Retrieved September 29, 2013, from http://www.express.co.uk/news/weird/86334/Primitive-lot-these-shoppers

BBC news. (2011). LSE lecturer Dr Satoshi Kanazawa tells of race blog "regret." *BBS News London.* Retrieved May 28, 2013, from http://www.bbc.co.uk/news/uk-england-london-14945110

BBC News. (2011). LSE investigates lecturer's blog over race row. *BBC News.* Retrieved May 28, 2013, from http://www.bbc.co.uk/news/uk-13452699

Bouton, K. (2010). "Delusions of Gender" Peels Away Popular Theories. *The New York Times website.* Retrieved May 28, 2013, from http://www.nytimes.com/2010/08/24/science/24scibks.html

Buss, David M., and Meston, C. M. (2009). Why do women have sex? (David Buss and Cindy Meston at CASW 2009). *YouTube.com.* Retrieved May 28, 2013, from http://www.youtube.com/watch?v=KA0sqg3EHm8&feature=youtu.be&t=5m10s

Buss, David Michael, and Schmitt, D. P. (2011). Evolutionary Psychology and Feminism. *Sex Roles, 64*(9-10), 768–787. doi:10.1007/s11199-011-9987-3

Castleman, M. (2010). Why Do People Have Sex? | Psychology Today. *Psychology Today.* Retrieved May 28, 2013, from http://www.psychologytoday.com/blog/all-about-sex/201011/why-do-people-have-sex

Clark, R., and Hatfield, E. (1989). Gender Differences in Receptivity to Sexual Offers. *Journal of Psychology and*

Human Sexuality, 2(1), 39–55.
doi:10.1300/J056v02n01_04

Clint, E. K., Sober, E., Garland Jr., T., and Rhodes, J. S. (2012).
Male Superiority in Spatial Navigation: Adaptation or Side
Effect? *The Quarterly Review of Biology, 87*(4), 289–313.
doi:10.1086/668168

Confer, J. C., Easton, J. A., Fleischman, D. S., Goetz, C. D.,
Lewis, D. M. G., Perilloux, C., and Buss, D. M. (2010).
Evolutionary psychology. Controversies, questions,
prospects, and limitations. *The American psychologist,
65*(2), 110–26. doi:10.1037/a0018413

Conley, T. D. (2011). Perceived proposer personality
characteristics and gender differences in acceptance of
casual sex offers. *Journal of Personality and Social
Psychology, 100*(2), 309–329. doi:10.1037/a0022152

Cosmides, L., and Tooby, J. (1997). Evolutionary Psychology: A
Primer. *Center for Evolutionary Psychology website.*
Retrieved September 29, 2013, from
http://www.cep.ucsb.edu/primer.html

Enserink, M. (2010a). Elsevier to Editor: Change Controversial
Journal or Resign. *Science Insider.* Retrieved September
29, 2013, from
http://news.sciencemag.org/2010/03/elsevier-editor-
change-controversial-journal-or-resign

Enserink, M. (2010b). Elsevier to Editor: Change Controversial
Journal or Resign. *ScienceInsider.* Retrieved May 28, 2013,
from
http://news.sciencemag.org/scienceinsider/2010/03/else
vier-to-editor-change-contro.html

Fink, B., Hugill, N., and Lange, B. P. (2012). Women's body
movements are a potential cue to ovulation. *Personality
and Individual Differences, 53*(6), 759–763. Retrieved from
http://www.sciencedirect.com/science/article/pii/S01918
6912002930

Finkel, E. J., and Eastwick, P. W. (2009). Arbitrary social norms
influence sex differences in romantic selectivity.
Psychological science, 20(10), 1290–5. doi:10.1111/j.1467-
9280.2009.02439.x

Fleischman, D. S. (2013). Sentientist. *Sentientist.com.* Retrieved
from http://sentientist.org/

Forbes, P. (2013). Paleofantasy: What Evolution Really Tells Us About Sex, Diet, and How We Live by Marlene Zuk—review. *The Guardian*. Retrieved May 28, 2013, from http://www.guardian.co.uk/books/2013/apr/24/paleofantasy-evolution-sex-diet-review

Gangestad, S. W., Thornhill, R., and Garver, C. E. (2002). Changes in women's sexual interests and their partners' mate-retention tactics across the menstrual cycle: evidence for shifting conflicts of interest. *Proceedings. Biological sciences / The Royal Society, 269*(1494), 975–82. doi:10.1098/rspb.2001.1952

Gendered Division Of Labor Gave Modern Humans Advantage Over Neanderthals. (2006). *ScienceDaily*. Retrieved May 28, 2013, from http://www.sciencedaily.com/releases/2006/12/061204123302.htm

George, A. (2013a). Marlene Zuk's Paleofantasy book: Diets and exercise based on ancient humans are a bad idea. *Slate*. Retrieved May 28, 2013, from http://www.slate.com/articles/health_and_science/new_scientist/2013/04/marlene_zuk_s_paleofantasy_book_diets_and_exercise_based_on_ancient_humans.html

George, A. (2013b). Should we aim to live like cavemen? *NewScientist*. Retrieved May 28, 2013, from http://www.newscientist.com/article/mg21729090.400-should-we-aim-to-live-like-cavemen.html

Goldacre, B. (2007). Imaginary numbers—Bad Science. *Bad Science*. Retrieved May 28, 2013, from http://www.badscience.net/2007/09/imaginary-numbers/

Gonzaga, G. C., Haselton, M. G., Smurda, J., Davies, M. sian, and Poore, J. C. (2008). Love, desire, and the suppression of thoughts of romantic alternatives☆. *Evolution and Human Behavior, 29*(2), 119–126. doi:10.1016/j.evolhumbehav.2007.11.003

Grammer, K., Renninger, L., and Fischer, B. (2004). Disco clothing, female sexual motivation, and relationship status: is she dressed to impress? *Journal of sex research, 41*(1), 66–74. doi:10.1080/00224490409552214

Guéguen, N. (2011). Effects of solicitor sex and attractiveness on receptivity to sexual offers: a field study. *Archives of sexual behavior, 40*(5), 915–9. doi:10.1007/s10508-011-9750-4

Guide for authors. (2012). *Medical Hypotheses.* Retrieved May 28, 2013, from http://www.elsevier.com/journals/medical-hypotheses/0306-9877/guide-for-authors

Hagen, E. (2004). Evolutionary Psychology FAQ. *University of California Santa Barbara Anthropology website.* Retrieved May 28, 2013, from http://www.anth.ucsb.edu/projects/human/epfaq/evpsychfaq_full.html#holocene

Halpern, D. F. (2010). How Neuromythologies Support Sex Role Stereotypes. *Science, 330*(6009), 1320–1321. doi:10.1126/science.1198057

Haselton, M. G., and Gildersleeve, K. (2011). Can Men Detect Ovulation? *Current Directions in Psychological Science, 20*(2), 87–92. doi:10.1177/0963721411402668

Haselton, Martie G, and Gangestad, S. W. (2006). Conditional expression of women's desires and men's mate guarding across the ovulatory cycle. *Hormones and behavior, 49*(4), 509–18. doi:10.1016/j.yhbeh.2005.10.006

Haselton, Martie G, Mortezaie, M., Pillsworth, E. G., Bleske-Rechek, A., and Frederick, D. A. (2007). Ovulatory shifts in human female ornamentation: near ovulation, women dress to impress. *Hormones and behavior, 51*(1), 40–5. doi:10.1016/j.yhbeh.2006.07.007

Havlicek, J., Dvorakova, R., Bartos, L., and Flegr, J. (2006). Non-Advertized does not Mean Concealed: Body Odour Changes across the Human Menstrual Cycle. *Ethology, 112*(1), 81–90. doi:10.1111/j.1439-0310.2006.01125.x

Holmes, D. (2012). MMU | Research Institute for Health and Social Change. *Manchester Metropolitan University website.* Retrieved May 28, 2013, from http://www.rihsc.mmu.ac.uk/staff/profile.php?surname=Holmes&name=David

Hurlbert, A. C., and Ling, Y. (2007). Biological components of sex differences in color preference. *Current biology : CB, 17*(16), R623–5. doi:10.1016/j.cub.2007.06.022

Kalant, H., Pinker, S., Kalow, W., and Gould, S. J. (1997). Evolutionary psychology: An exchange. *New York Review of Books*, (44), 55–56.

Kaplan, M. (2009, June 2). Role reversal undermines speed-dating theories. *Nature*. Nature Publishing Group. doi:10.1038/news.2009.537

Khamsi, R. (2007). Women may be hardwired to prefer pink. *NewScientist.com*. Retrieved from http://www.newscientist.com/article/dn12512-women-may-be-hardwired-to-prefer-pink.html#.UlfRN1Cfh1F

Kirchengast, S., and Gartner, M. (2002). Changes in fat distribution (WHR) and body weight across the menstrual cycle. *Collegium antropologicum, 26 Suppl*, 47–57. Retrieved from http://www.ncbi.nlm.nih.gov/pubmed/12674835

Krug, R., Mölle, M., Fehm, H. L., and Born, J. (1999). Variations across the menstrual cycle in EEG activity during thinking and mental relaxation. *Journal of Psychophysiology*, (13), 163–172.

Kruger, D., and Byker, D. (2009). EVOLVED FORAGING PSYCHOLOGY UNDERLIES SEX DIFFERENCES IN SHOPING EXPERIENCES AND BEHAVIORS. *Journal of Social, Evolutionary, and Cultural Psychology*, 3(4), 328–342.

Kuhn, S. L., and Stiner, M. C. (2006). What's a mother to do? A hypothesis about the division of labor and modern human origins. *Current Anthropology*. Retrieved September 29, 2013, from http://www.academia.edu/968719/Whats_a_mother_to_d o_A_hypothesis_about_the_division_of_labor_and_modern_ human_origins

Kurzban, R. (2013a). Mind Design | Psychology Today. *Psychology Today*. Retrieved May 28, 2013, from http://www.psychologytoday.com/blog/mind-design

Kurzban, R. (2013b). Evolutionary Psychology blog. *Evolutionary Psychology*. Retrieved from http://www.epjournal.net/blog/

Kuukasjarvi, S. (2004). Attractiveness of women's body odors over the menstrual cycle: the role of oral contraceptives and receiver sex. *Behavioral Ecology*, 15(4), 579–584. doi:10.1093/beheco/arh050

Leach, B. (2009). Shopping is "throwback to days of cavewomen." *The Telegraph.* Retrieved May 27, 2013, from http://www.telegraph.co.uk/news/newstopics/howaboutthat/4803286/Shopping-is-throwback-to-days-of-cavewomen.html

Lombardo, W. K., Cretser, G. A., and Roesch, S. C. (2002). For Crying Out Loud — The Differences Persist into the '90s, *45*(October 2001), 529–547.

Lovgren, S. (2006). Sex-Based Roles Gave Modern Humans an Edge, Study Says. *National Geographic website.* Retrieved May 28, 2013, from http://news.nationalgeographic.com/news/2006/12/061207-sex-humans.html

Lucas, M. (2012). Cordelia Fine, Delusions of Gender: How our Minds, Society, and Neurosexism Create Difference. *Society, 49*(2), 199–202. doi:10.1007/s12115-011-9527-3

Manning, J. T., Scutt, D., Whitehouse, G. H., Leinster, S. J., and Walton, J. M. (1996). Asymmetry and the menstrual cycle in women. *Ethology and Sociobiology,* (17), 129–143. Retrieved from http://www.journals.elsevierhealth.com/periodicals/ensol d/article/0162-3095(96)00001-5/references

McDonald, M. M., Asher, B. D., Kerr, N. L., and Navarrete, C. D. (2011). Fertility and intergroup bias in racial and minimal-group contexts: evidence for shared architecture. *Psychological science, 22*(7), 860–5. doi:10.1177/0956797611410985

Meston, C. M., and Buss, D. M. (2007). Why humans have sex. *Archives of sexual behavior, 36*(4), 477–507. doi:10.1007/s10508-007-9175-2

Miller, G., Tybur, J., and Jordan, B. (2007). Ovulatory cycle effects on tip earnings by lap dancers: economic evidence for human estrus? *Evolution and Human Behavior, 28*(6), 375–381. doi:10.1016/j.evolhumbehav.2007.06.002

Miller, L. (2013). "Paleofantasy": Stone Age delusions. *Salon.com.* Retrieved from http://www.salon.com/2013/03/10/paleofantasy_stone_age_delusions/

Miller, S., and Maner, J. (2010). Evolution and relationship maintenance: Fertility cues lead committed men to devalue

relationship alternatives. *Journal of Experimental Social Psychology, 46*(6), 1081–1084.
doi:10.1016/j.jesp.2010.07.004

Navarrete, C. D., Fessler, D. M. T., Fleischman, D. S., and Geyer, J. (2009). Race bias tracks conception risk across the menstrual cycle. *Psychological science, 20*(6), 661–5. doi:10.1111/j.1467-9280.2009.02352.x

Navarrete, C. D., McDonald, M. M., Mott, M. L., Cesario, J., and Sapolsky, R. (2010). Fertility and race perception predict voter preference for Barack Obama. *Evolution and Human Behavior, 31*(6), 394–399.
doi:10.1016/j.evolhumbehav.2010.05.002

Nuwer, R. (2013). Men Are Better Navigators Than Women, But Not Because of Evolution. *The Smithsonian.com.* Retrieved May 28, 2013, from
http://blogs.smithsonianmag.com/smartnews/2013/02/men-are-better-navigators-than-women-but-not-because-of-evolution/

Penke, L., Borsboom, D., Johnson, W., Kievit, R. A., Ploeger, A., and Wicherts, J. M. (2011). Evolutionary Psychology and Intelligence Research Cannot Be Integrated the Way Kanazawa (2010) Suggested. *American Psychology, 66*(9), 916–7. Retrieved from
http://www.larspenke.eu/pdfs/Penke_et_al_in_press_-_Kanazawa_commentary.pdf

Pillsworth, E. G., and Haselton, M. G. (2006). Male sexual attractiveness predicts differential ovulatory shifts in female extra-pair attraction and male mate retention. *Evolution and Human Behavior, 27*(4), 247–258. doi:10.1016/j.evolhumbehav.2005.10.002

Pinker, S. (1997). *How the Mind Works* (p. 660). New York: W W Norton and Co Inc. Retrieved from
http://www.amazon.com/How-Mind-Works-Steven-Pinker/dp/0393045358

Pinker, S. (2012). Steven Pinker's correspondence with journalist Dan Slater, 2012. Retrieved May 28, 2013, from
http://stevenpinker.com/files/pinker/files/pinker_corredspondence_with_dan_slater.pdf

Provost, M. P., Quinsey, V. L., and Troje, N. F. (2008). Differences in gait across the menstrual cycle and their attractiveness to men. *Archives of sexual behavior*, *37*(4), 598–604. doi:10.1007/s10508-007-9219-7

Roberts, S. C., Havlicek, J., Flegr, J., Hruskova, M., Little, A. C., Jones, B. C., ... Petrie, M. (2004). Female facial attractiveness increases during the fertile phase of the menstrual cycle. *Proceedings. Biological sciences / The Royal Society*, *271 Suppl* (Suppl_5), S270–2. doi:10.1098/rsbl.2004.0174

SD-agencies. (2009). Shopping is "throwback to days of cave women:" study. *Shenzhen Daily*. Retrieved September 29, 2013, from http://szdaily.sznews.com/html/2009-02/26/content_528445.htm#

Self Magazine. (2011). TresSugar/SELF Magazine Casual Sex Survey. *Self Magazine*. Retrieved September 30, 2013, from http://www.tressugar.com/TresSugarSELF-Magazine-Casual-Sex-Survey-Results-16742092?image_nid=16742092

Singh, D., and Bronstad, P. M. (2001). Female body odour is a potential cue to ovulation. *Proceedings. Biological sciences / The Royal Society*, *268*(1469), 797–801. doi:10.1098/rspb.2001.1589

Slater, D. (2013). Darwin Was Wrong About Dating. *New York Times website*. Retrieved May 28, 2013, from http://www.nytimes.com/2013/01/13/opinion/sunday/darwin-was-wrong-about-dating.html?pagewanted=all

Switek, B. (2010). The earliest known pelican reveals 30 million years of evolutionary stasis in beak morphology. *Wired.com*, *152*(1), 15–20. doi:10.1007/s10336-010-0537-5

Symonds, C. S., Gallagher, P., Thompson, J. M., and Young, A. H. (2004). Effects of the menstrual cycle on mood, neurocognitive and neuroendocrine function in healthy premenopausal women. *Psychological medicine*, *34*(1), 93–102. Retrieved from http://www.ncbi.nlm.nih.gov/pubmed/14971630

Symons, D. (2000). Regarding V.S. Ramachandran. *Center for Evolutionary Psychology website*. Retrieved May 28, 2013, from http://www.cep.ucsb.edu/ramachandran.html

Telegraph Media Group. (2007). Jessica Alba has the perfect wiggle, study says. *Telegraph*. Retrieved May 28, 2013,

from
http://www.telegraph.co.uk/news/uknews/1561306/Jess
ica-Alba-has-the-perfect-wiggle-study-says.html

Thornhill, R., and Palmer, C. T. (2001). *A Natural History of Rape: Biological Bases of Sexual Coercion* (p. 272). MIT Press. Retrieved from
http://books.google.com/books?hl=en&lr=&id=xH6v-nB6EegC&pgis=1

Tooby, J., and Cosmides, L. (2000). Response to Coyne. *Center for Evolutionary Psychology website.* Retrieved May 28, 2013, from http://www.cep.ucsb.edu/tnr.html

Van Hemert, D. A., van de Vijver, F. J. R., and Vingerhoets, A. J. J. M. (2011). Culture and Crying: Prevalences and Gender Differences. *Cross-Cultural Research, 45*(4), 399–431. doi:10.1177/1069397111404519

Villotte, S., Churchill, S. E., Dutour, O. J., and Henry-Gambier, D. (2010). Subsistence activities and the sexual division of labor in the European Upper Paleolithic and Mesolithic: evidence from upper limb enthesopathies. *Journal of Human Evolution, 59*(1), 35–43. doi:10.1016/j.jhevol.2010.02.001

Ward, A. F. (2012). Scientists Probe Human Nature--and Discover We Are Good, After All. *ScientificAmerican.com.* Retrieved May 28, 2013, from
http://www.scientificamerican.com/article.cfm?id=scientis
ts-probe-human-nature-and-discover-we-are-good-after-all

Wayman, E. (2012). When Did the Human Mind Evolve to What It is Today? *The Smithsonian.* Retrieved May 28, 2013, from http://www.smithsonianmag.com/science-nature/When-Did-the-Human-Mind-Evolve-to-What-It-is-Today-160374925.html

Weierstall, R., Schauer, M., and Elbert, T. (2013). Psychology of War Helps to Explain Atrocities. *Scientific American.* Retrieved May 28, 2013, from
http://www.scientificamerican.com/article.cfm?id=psychol
ogy-war-helps-explain-atrocities

Wilson, D. S. (2013). Evolution: This View of Life Magazine. *Evolution: This View of Life.* Retrieved from http://www.thisviewoflife.com/index.php

Zimmer, C. (2013). Of Men, Navigation, and Zits—Phenomena. *National Geographic website.* Retrieved May 28, 2013, from http://phenomena.nationalgeographic.com/2012/12/28/of-men-navigation-and-zits/

Zuk, M. (2013). *Paleofantasy: What Evolution Really Tells Us about Sex, Diet, and How We Live.* W. W. Norton and Company. Retrieved from http://www.amazon.com/Paleofantasy-Evolution-Really-Tells-ebook/dp/B007Q6XM1A

Author Biographies

Russell Blackford, *The Hellfire Club*:

Russell Blackford is an Australian philosopher, literary critic, editor, and author. He is a Conjoint Lecturer in the School of Humanities and Social Science, University of Newcastle, NSW. His many books include, most recently, *50 Voices of Disbelief: Why We Are Atheists* (co-edited with Udo Schüklenk; Wiley-Blackwell, 2009), *Freedom of Religion and the Secular State* (Wiley-Blackwell, 2012), *50 Great Myths About Atheism* (co-authored with Udo Schüklenk; Wiley-Blackwell, 2013), and *Humanity Enhanced: Genetic Choice and the Challenge for Liberal Democracies* (MIT Press, 2014).

He is a prolific essayist and commentator whose interests include legal and political philosophy, philosophical bioethics, philosophy of religion, and debates involving visions of the human future. Dr. Blackford is a Fellow of the Institute for Ethics and Emerging Technologies, and Editor-in-Chief of *The Journal of Evolution and Technology*. He blogs at *Talking Philosophy* in addition to *Skeptic Ink*.

Rebecca Bradley, *The Lateral Truth*:

Rebecca Bradley was a congenital atheist born into a fundamentalist family in Vancouver, Canada. A Nile Valley archaeologist by training and a writer by inclination, she received her PhD from Cambridge University and lived abroad for much of her adult life, returning to Canada to teach archaeology in 1996. Recently, she retired with her husband to an acreage in the Kootenays of British Columbia, to write, raise chickens, and be ruled by her cats. She is the author of several novels of speculative fiction and many short stories, including a volume of blasphemous biblical variations entitled *The Lateral Truth*.

267

Edward Clint, *Incredulous*:

Edward Clint co-founded the *Skeptic Ink Network*, an online salon parsing issues in science, skepticism, secularism and other topics for public consumption and discussion. He is a bioanthropology graduate student at UCLA studying evolutionary psychology, a subject which he writes about at his own *Skeptic Ink* channel, *Incredulous*. Edward is also a former United States Air Force airman and award-winning secular student group leader.

Peter Ferguson, *Humanisticus*:

Peter Ferguson holds a BA in History and Classics and an MA in Classics, both from the National University of Ireland, Galway. He is a member of Atheist Ireland and the Humanist Association of Ireland. He founded the Humanist Atheist Society in his university and was the Auditor for the inaugural year when the society was awarded "The Food for Thought" award for adding to the intellectual diversity of the university and holding the most thought-provoking events. Peter's current focus is current affairs related to secularism; however, he also has a keen interest in the early development of Christianity and how this influenced society.

John W. Loftus, *Debunking Christianity*:

John W. Loftus is a former Christian minister and apologist with M.A., M.Div., and Th.M. degrees in Philosophy, Theology, and the Philosophy of Religion, the last of which was earned under William Lane Craig. He is the author of *Why I Became an Atheist: A Former Preacher Rejects Christianity*, and *The Outsider Test for Faith*. He edited *The Christian Delusion*, and *The End of Christianity*. He has also co-authored, with Dr. Randal Rauser, *God or Godless?*

Caleb Lack, *Great Plains Skeptic*:

Caleb W. Lack, Ph.D. is a clinical psychologist and professor at the University of Central Oklahoma. The author of over three dozen scientific articles, books, and book chapters, Dr. Lack's clinical research and practice are focused on the anxiety and obsessive-compulsive disorders. He also teaches courses on critical thinking that produced the edited text *Science, Pseudoscience, & Critical Thinking* and a series of documentaries on pseudoscience and superstition in Oklahoma. He is the *Great Plains Skeptic* on the *Skeptic Ink Network*, is on CFI's Speaker's Bureau, and collaborates with both the James Randi Educational Foundation and the Skeptics Society.

Kevin McCarthy, *Smilodon's Retreat*:

Kevin is a former high school science teacher who now works in the assessment industry. He has taught and studied most main branches of physical sciences, but he especially loves evolution, taxonomy, high-energy physics, and cosmology. He is also interested in studying the creationism movement.

Kevin has written for several blogs, including a guest stint at *Scientopia*, and is now the *Skeptic Ink* blog *Smilodon's Retreat*. He has also published articles with several publishers, including the European Space Agency on Innovative Ideas from Science Fiction.

When not writing or thinking about science, he can be found playing games at the local game stores and online.

Maria Maltseva, *Skeptically Left*:

Maria Maltseva is an attorney, writer, editor, and part-time piano teacher living in Seattle, Washington. She has more college degrees than she can count (including BAs in communications, psychology, and Russian language, as well as a juris doctor and a Master's degree in law), and if there were only one thing she could do with her life, she would spend it studying everything she has not yet had the chance to learn. In her free time, Maria enjoys roaming the internet, exploring this pale blue dot, and getting lost in her dreams. Her favorite things include snow, science, music, and art, though not necessarily in that order. She's currently studying to get her sport pilot's license and, one day, she'd like to move to New York. Maria is an active advocate for LGBTQ rights and also volunteers to help low income people avoid mortgage foreclosure. She has participated in several piano competitions and won first place for playing Gershwin's Rhapsody in Blue at the Eastside Music Festival in Bellevue, Washington. Aside from that, she's fond of poetry, photography, skiing, writing, and civil online debate.

David Osorio, *Avant Garde*:

David Osorio, a Colombian journalist, always disliked religions' antagonism toward freedom. He became an Atheist after reading Christopher Hitchens' *God Is Not Great*, and not long after that, he became a full-time skeptic, with contempt for pseudosciences and conspiracy theories. By 2012 he, and others, founded the Bogotá Atheists and Agnostics Association, of which he is now the spokesperson. He blogs both in Spanish (at http://de-avanzada.blogspot.com/) and English (at http://skepticink.com/avant-garde/) about religion, science, atheism, politics, skepticism and crazy stuff.

Jonathan MS Pearce, *A Tippling Philosopher*:

Holding a degree from the University of Leeds, a PGCE from the University of St Mary's, Twickenham and a Masters in Philosophy from the University of Wales, Pearce has written *Free Will? An investigation into whether we have free will or whether he was always going to write this book* as well as *The Little Book Of Unholy Questions* and *The Nativity: A*

Critical Examination. His latest work is as editor of an anthology of deconversion accounts entitled *Beyond an Absence of Faith.* He blogs under the name *A Tippling Philosopher* (skepticink.com/tippling), being a big advocate of casual philosophy and a core member of The Tippling Philosophers, a pub philosophy and theology group. Working as a teacher, he lives in Hampshire, UK with his partner and twin boys. Follow him on Twitter: @ATipplingPhilo.

Staks Rosch, *Dangerous Talk*:

Staks Rosch is an agnostic atheist, a Secular Humanist, a skeptic, a philosopher, stay-at-home dad, and a Jedi. He is a former board member of the Freethought Society of Greater Philadelphia and currently serves as the head of the Philadelphia Coalition of Reason. He writes as both the Philadelphia Atheism Examiner and the National Atheism Examiner at Examiner.com and is a Huffington Post Contributor.

Rosch appeared as a panelist on the CN8 television show *"It's Your Call with Lynn Doyle"* arguing against Intelligent Design in the classroom. He was a guest on the Christian radio show, *"Bob Enyart Live"* to discuss atheism.

Dangerous Talk started as a college radio show dealing with politics, social issues, and atheism. The show then moved to a local AM Radio station. It eventually became a daily blog before becoming a part of the *Skeptic Ink Network.*

Follow Staks Rosch on Twitter and Facebook @DangerousTalk.

Jacques Rousseau, *Towards A Free Society*:

Jacques Rousseau is a lecturer in critical thinking and ethics at the University of Cape Town, and the founder and Chair of the Free Society Institute, a South African non-governmental organization dedicated to defending free speech and the secular viewpoint against threats presented by religion, bad science and other forms of irrationality. He writes a weekly column for the online newspaper the *Daily Maverick*, blogs at http://synapses.co.za, and is @JacquesR on Twitter.

Hugo Schmidt, *The Prussian*:

Hugo Schmidt is a molecular and computational biologist, finishing his Ph.D. at the University of Cambridge. The so-called "Muhammad" cartoon riots dragged him reluctantly into the public sphere, and he has written and published on both sides of the Atlantic from an Objectivist perspective, though he would far rather be left alone at the lab and in the library.

Zachary Sloss, *Notung*:

A life-long non-believer, Zachary Sloss developed a keen interest in atheism and skepticism in his early twenties, and completed an MA in philosophy at the University of York in 2012. Zachary also has a bachelor's degree and a graduate diploma in music, and has played trombone for various European orchestras. He currently lives in Taiwan, and enjoys listening to Wagner and watching Manchester United.

CPSIA information can be obtained
at www.ICGtesting.com
Printed in the USA
LVOW13s0418080817
544215LV00013B/122/P